Manual of Electrical Contracting

by Bob Dries

Craftsman Book Company
6058 Corte Del Cedro, Carlsbad, CA 92008

Acknowledgements

The author would like to thank the following for their assistance in preparing the text:

Amprobe Instrument Co.,   P.O. Box 329, 630 Merrick Rd., Lynbrook, NY 11563

Construction Bookstore, Inc.   P.O. Box 717, 1830 N.E. 2nd St., Gainesville, FL 32602

Dries-Jacques Associates, Mechanical Engineers   1600 Verona St., Middleton, WI 53562

Food and Energy Council   909 University Ave., Columbia, MO 65201

The Illuminating Engineering Society   345 E. 47th St., New York, NY 10017

Leviton Manufacturing Co.   59-25 Little Neck Parkway, Little Neck, NY 11362

The National Electrical Contractors' Association   7315 Wisconsin Ave., Washington, D.C. 20014

The National Electrical Manufacturers' Association   2101 L Street, N.W., Washington, D.C. 20037

Square D Company   Executive Plaza, Palatine, IL 60067

Thomas Industries Inc.   207 E. Broadway, Louisville, KY 40202

This manual is based on the author's 15 years of experience and training in the construction industry and conforms with rules and procedures that are generally accepted by electricians and electrical contractors. But neither the author, publisher, nor any authority cited in this book can accept responsibility for information presented here which is inconsistent with standards used in your community. There is no single authority that applies without exception in every state, county and city. What's acceptable practice in many areas may be prohibited or restricted in others. This book is not a substitute for the code, ordinances, regulations, standards and statutes that have been adopted in the area or areas where you do electrical work. Get a copy of your local code and rely on it as final authority for what will meet local requirements.

Library of Congress Cataloging in Publication Data

Dries, Bob.
  Manual of electrical contracting.

  Includes index.
  1. Electric contracting—Handbooks, manuals, etc.
I. Title
TK441.D74   1983   621.319'24'068   83-7232
ISBN 0-910460-33-7

Illustrations by Richard Abernethy and Linda Hokin.
Technical advice by William G. Dries, Engineering Specialist, University of Wisconsin, Madison, WI.

Contents

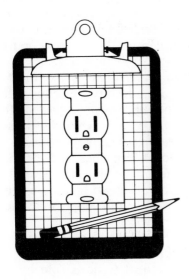

Chapter 1:

Codes, Permits and Utilities

The electrician's trade is demanding and exacting work. It requires both knowledge and a high degree of skill. No wonder electricians are among the most highly paid of all construction tradesmen.

This manual is a guide to the essential knowledge and basic skills every electrician and electrical contractor needs. Of course, there's no substitute for on-the-job experience, but this book will speed your learning many times over.

Most chapters cover electrical materials, tools, planning and doing the work. Others explain the business of electrical contracting: getting started, estimating costs, and staying profitable.

Throughout this book we refer to the guide for all electrical installation, the National Electrical Code. In some sections we explain what the code requires. You should understand that this manual isn't a substitute for the code. You and every electrical contractor need a copy of the NEC.

This is a practical book. It isn't an abstract text. Neither is it a handbook for electrical engineers. Theory will be kept to a minimum. But every electrician needs to know more about electricity than carpenters need to know about trees. You can't follow the code and use accepted practice if you don't understand the physical properties and fundamental laws of electricity. That's why Chapter 2 covers the basic electrical principles.

This chapter covers the rules government has established for electrical contractors: the code, permits, and licensing.

The National Electrical Code
For over 70 years the National Electrical Code has been the accepted standard for electrical work in both private and public construction throughout the United States. The first version of the NEC was drawn up in 1897 by electrical, architectural and insurance firms. Since 1911, the National Fire Protection Association has sponsored the code. The NFPA has the assistance of the American National Standards Institute and special advisory committees in producing each revision.

The code that the NFPA publishes every three years has no binding authority by itself. It's given authority when adopted as a regulation by states, counties and city governments. Each jurisdiction makes the code apply to work done under its authority. Some units of government adopt a slightly different version of the code or add their own sections. These revisions are usually rules local inspectors feel are needed for their area. Some jurisdictions enforce older versions of the code because they have not adopted the latest amendments published by the NFPA.

Be sure you know what version of the NEC is being enforced where you do work and be familiar with any revisions that apply.

Don't assume that all an electrical contractor needs to know is the code. It won't tell you how to do the work. It doesn't guarantee efficiency, economy, convenience or provide for future building needs. These are your goals as a contractor or installer. You need to know much more than the code to be a good electrical contractor.

Some code provisions seem unnecessary or even a downright nuisance. But don't view the code as a hindrance rather than an aid. Realize that the code is your best guide. It may make your job a little more challenging. But it also protects you in several ways.

First, it practically guarantees safe wiring practice and proper installation. Second, if you've done what the code requires, you should have no trouble defending yourself against a claim of negligence. Finally, following the code exactly will prevent costly "do-overs" when the inspector makes his on-site inspection.

Here's another tip. For these same reasons, it's good practice to use materials approved by Underwriter's Laboratories. In some cases, UL-approved materials are required by job specifications or NEC regulations. But many types of material are available without the UL seal of approval. Avoid these when possible.

The NEC is reasonably well indexed. It's relatively easy to look up the standards for the job you are handling. Your responsibility is to be aware of all regulations that apply to your job. If you're unclear about the meaning of any code provisions, your building inspector will be happy to explain exactly what that section means to him—especially if you inquire *before* starting the work.

It's possible to have work approved that does not comply with the NEC. States, cities and counties have the right to approve anything they want to approve. But you're going to need some good arguments to back up your claim for an exception. You need to show that the way you want to do the work is as safe or safer in this case than if you followed the NEC. If you get approval of an exception, *get it in writing* before beginning work. Ask the inspector to initial his approval on your letter requesting the exception.

Licensing and Permits

Many cities and counties place restrictions on who may install electrical materials. Generally, the more populated an area, the stricter the requirements. In most larger cities, for example, only a licensed electrician can install wiring in new homes. Electricians are licensed by the city after passing a test. Electrical work permits are issued only to these licensed electricians. An unlicensed person may, however, be permitted to work alongside the licensed electrician on the job itself.

For residences already occupied, provisions may be different. Sometimes electrical remodeling work can be done by the owner, but it must pass an inspection when completed.

In some communities the owner can wire a new residence but must have a licensed electrician con-

nect the circuits to the electrical panel. In other areas the owner can do all the inside wiring. But the outside service must be done by the electrical utility. In some rural areas the owner can do all the work short of bringing in the service drop from the power line.

Plans for *commercial work* usually must be approved by a state, county or city authority. Sometimes several approvals are needed. In some cases these plans must be drawn under the supervision of an engineer licensed by the state.

Know what local licensing and permit procedures apply *before* starting work. Always get exceptions and approvals in writing. State your position clearly. Be friendly and cooperative and you'll be treated the same way.

Where to get Codes and Permits

The National Electrical Code is sold by the National Fire Protection Association, 470 Atlantic Avenue, Boston, Massachusetts 02210 and by many larger bookstores. Code publication dates are 1981, 1984, 1987 and 1990.

The office of your building inspector will furnish information about other codes and licensing requirements enforced in that jurisdiction. Your local inspector can also supply the address of the state inspection office, if needed.

Working with the Utility and Supplier

On many jobs you'll have to coordinate your work with the electrical utility supplying the power. Most utilities reserve the right to select the types of material and equipment served by their lines. This is especially true for underground service cables. Even a panel that is both UL-approved and meets code requirements won't necessarily be accepted by the utility. Check this out when planning the job. Don't buy materials until you're satisfied that they are acceptable to the utility.

Many utilities supply guides that explain their requirements. Some also provide general instruction materials. Most have field representatives who will meet with the electrician at the job site if necessary. Some companies also have specialists to aid in planning present and future electrical needs for a business or dwelling.

Another place to get help is your electrical supplier. Whether the installation is a one-time job or one of many planned, it's cheapest to work through a wholesaler. The supplier should be willing to do more than sell material. Find one with experienced salespeople. They can help you select the proper materials or suggest ways to cut costs while still doing a good job.

If a problem does come up, a good wholesaler can usually put you in touch with someone who has had similar difficulties. A single phone call may save you thousands of dollars.

One word of caution: No matter how good their intentions, suppliers and other installers didn't write the code and don't enforce it. Those responsibilities belong to the NFPA and your inspector. Only your inspector can tell you what will pass inspection.

Safety

Contrary to what you might expect, the electrical trade is one of the safest, if not *the* safest, of the construction trades. That's why worker's compensation insurance rates for electrical tradesmen are usually lower than for any other construction trade. You'll find that most electricians take real pride in working safely. Some even would claim, only partly in jest, that as a group, electricians are too smart to take chances with safety.

A safe job doesn't just happen. It's the result of a deliberate plan that results in safe working conditions and a safety-conscious attitude for every man on the job. Here are some trademarks of a safe job: The right tools are available and in good working order. The work area is cleaned of accumulated debris regularly. Ladders and scaffolding are adequate for the work at hand and are in good condition. Every electrician and apprentice is aware that safety is part of his job.

Safety doesn't stop with protecting your own crews. Usually you'll have to provide temporary power for the other tradesmen (Figure 1-1). The NEC requires that these circuits be ground-fault, circuit-interrupter types. This is explained in Chapter 2. Be sure these circuits are sturdily constructed and well protected against hard use. They'll get more than their share.

All electrical work should be done on dead circuits when possible. The best way is to shut off the main. If this isn't possible, cut the circuit to the work area. Remove the fuses and place them away from the service panel. In a breaker panel, put up a hand-written sign that explains why the switches are off. Better yet, close and lock the cover. That prevents some obliging individual working on another part of the site from trying to correct a "power failure" when you're up to your elbows in bare wires.

Work on "hot" circuits only when there's no alternative. When this is necessary, only an experienced electrician should do the work. Pay attention to where you're standing. Standing in water or on a damp surface makes you an excellent con-

A Weather-proof Temporary Service for
Construction Work
Figure 1-1

ductor to ground. Use insulated tools and gloves. Be careful where you put your hands and fingers. Work only on one side of the circuit at a time. Make sure critical wires are secure to keep them from popping out and touching another wire or tool. Work cautiously, and think out each step.

On most new work it's fairly easy to keep track of the circuit connections. But on remodeling jobs you may find a bewildering number of wires in a junction box. Know where each wire goes before making any connection. In older work, wire markings and colors may not always be correct. Test each conductor and establish its true identity. Don't assume that because you've pulled a switch the circuit is dead. Double-check with a test lamp or voltmeter. This is essential on existing installations and is good practice on all jobs.

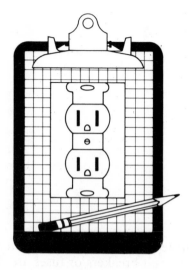

Chapter 2:

Basic Electrical Principles

This chapter explains the physical laws of electricity and the calculations you'll need as an electrical contractor. If you've been working as a licensed electrician for several years, you probably know everything in this chapter already. If you need a refresher on the basics, this chapter summarizes most of the essential electrical principles.

Most calculations an electrician has to handle require no more than simple addition, subtraction, multiplication or division. Some numbers will have to be squared (multiplied by itself). For others you'll need to find the square-root (the reverse process of squaring). A hand-held calculator selling for less than $20 will do this figuring quickly and accurately. It will also help in preparing bids and estimates.

Electrical Systems and Electrons
An electrical system is a series of mechanical connections that allows electrical energy to move from where it's generated to where it's needed. Electricity is the movement of *electrons*, the negative-charged particles of atoms. When force is applied to certain atoms, electrons can be made to move from their patterned movements into a path of lesser resistance.

Insulators and Conductors
Since all matter is made up of atoms, electron flow could occur in any material. But some materials give up electrons easily while other do so with great difficulty. Materials which give up electrons grudgingly are known as *insulators*. Examples are wood, glass and plastic. Materials which provide easy

electron flow are known as *conductors*. Copper, water and steel are examples. The word conductor, when used in electrical code books or instruction books, usually refers to the wires in the system. Electron flow in a conductor is *current*.

Circuits
A conductor or series of conductors installed in a system to allow the flow of current is called a *circuit*. A circuit consists of a voltage source, one or more loads which use the current provided by the voltage source, and the conductors used to carry the current.

Circuits are divided into two categories—series and parallel, each with its own characteristics and uses. They will be explained further in another section.

Electricians sometimes speak of two- and three-wire circuits. They're referring to a 115-volt or 230-volt circuit. Circuits also have *internal* and *external* resistance. These terms refer to the resistance in the circuit. Internal resistance is found at the voltage source. External resistance is in the connecting wires between the source and the load and in the load itself.

Amperage and Voltage
The flow of electrons through a conductor is electrical *current*. Current is measured in *amperes* or *amps*. The capacity of a conductor to carry current is called its *ampacity*.

The force that causes electrons to move through a conductor is called the *electromotive force* or *emf*

Name	Unit of Measure	Abbreviation
Current	Ampere, amp	A, a, I
Potential difference, Electromotive force	Voltage, volt	V, v, E
Resistance	Ohm	R, Ω
Power	Wattage, watt	W, w, P
# x 1000	Kilo-	K, k
# ÷ 1000	Milli-	m

Names, Units and Abbreviations
Table 2-1

of the circuit. It's also called the *potential difference*. Potential difference, also abbreviated "emf," is measured in *voltage* or *volts* (Table 2-1).

The name "potential difference" suggests that there is an imbalance in a circuit which causes current flow. In wiring diagrams this is shown by labeling one side of the power source "positive" and the other "negative." Current flows from the negative to the positive. Current and potential difference are generally spoken of as amps and volts.

A good way to understand the difference between voltage and amperage is to visualize water flowing through a garden hose. Current in the circuit is like water in the hose. It is measured in amperage rather than gallons per minute. But both the water and current can be measured in volume and do work for you. However, all the water in the world won't help you water your lawn if it stays in the hose. Water pressure, from a pump or gravity, pushes the water through the hose. Voltage does the same job in the electrical circuit. You need both pressure and volume to water your lawn. In an electrical circuit you need both volts and amps to run your toaster. Carrying the comparison a bit further, the hose is like the conductor in the electrical circuit.

Voltages Used
Voltages used in this text will be 115 and 230 volts. They were chosen to conform with voltages used in the NEC. Since there's a 10% leeway in voltage applications, these numbers can also serve as averages for voltages commonly used—110, 115, 117, 120, 220, 230, and 240. Calculations requiring the exact voltages will be noted.

It may be helpful to remember that, generally, as the voltage in a system increases, NEC regulations become more detailed. The NEC often divides systems according to the following:

- 50 volts or less
- 50 to 150 volts
- 150 to 300 volts
- 300 to 600 volts
- Over 600 volts

Resistance and Heat
Friction reduces the movement of electrons through a conductor. This is called *resistance*. Resistance is measured in units called *ohms*.

One important product of resistance is heat. Heat is used in many ways in a circuit. Toasters, electric heaters and stoves are all examples of heat developed through controlled resistance. Heat is also used to protect the system when it *shorts*, or malfunctions. Circuit breakers or fuses are designed to withstand only a certain amount of current. When there's a short, the uncontrolled current flow causes heat build-up. This trips the breaker or melts the fuse, protecting the circuit. Without the fuse breaker, the heat could destroy the wire and ignite the insulation.

All parts of an electrical system have limits on the amount of heat and resistance they can withstand. One important function of the NEC is to ensure that materials and equipment can withstand the loads (heat) they are intended to carry.

Power and Wattage
The rate of doing work through moving or changing energy is called *power*. Electrical power is measured in *wattage* or *watts*.

Units of Measurement
The value of volts, amperes, watts and ohms is usually stated directly. For example, we may speak of a 20 amp, 115-volt circuit having 6 ohms of resistance.

Occasionally, the prefix *kilo* is attached to a unit. A kilo multiplies the face value of a number by 1,000. So, 5,000 watts may be expressed as 5 kilowatts.

The term used to show values smaller than one is *milli*. Its value is a thousandth part of the whole unit. For example, if a small fan draws 0.5 amps, another way of saying it would be 500 milliamps.

These terms are often used on the selection boards of test instruments and on identifying tags, bands and *schematics* (wiring diagrams) of materials and equipment.

Watt-hours

A unit used to measure power over a period of time is the *watt-hour*. The utility company uses watt-hour meters or kilowatt-hour meters to measure the amount of current used by a customer. Meters are usually located at the service installation or, in commercial systems, at the distribution center. For newer residences they're located on the outside of the house. In some apartments, older buildings and commercial units they're often inside, next to the main service.

Horsepower

Since electrical power often does mechanical work, it is useful to relate them with a calculation.

Horsepower is a unit of mechanical power equal to 746 watts of electrical power. No motor is 100 percent efficient. For motors of 1 horsepower or larger, most experienced electricians estimate 1,000 watts of power consumption for each unit of horsepower. For fractional horsepower motors, estimate 1,200 watts for each horsepower output. For example, a 1/4-horsepower motor operating for two hours at 300 watts per hour would use about 600 watt-hours, or about 0.6 kilowatt-hours. A 2-horsepower motor operating for two hours at 2,000 watts per hour would use about 4,000 watt-hours or 4 kilowatt-hours. More information on calculating motor loads is given in Chapter 6.

Ohm's Law

Ohm's Law is the most fundamental law of electricity. It relates voltage, amperage and resistance. Ohm's Law states that current is equal to voltage divided by resistance.

$$\text{Current} = \frac{\text{Voltage}}{\text{Resistance}}$$

OR

$$\text{Amperes} = \frac{\text{Volts}}{\text{Ohms}}$$

OR

$$I = \frac{E}{R}$$

By substituting and cross-multiplying, we can find the resistance in a common 20 amp, 115 volt household circuit.

$$20\ A = \frac{115\ V}{R}$$

$$20 \times R = 115$$

$$R = \frac{115}{20}$$

$$R = 5.75\ \text{ohms}$$

Watt's Law

Watt's Law relates voltage and amperage to power. It is also called the *Power Law*. Watt's Law states that power equals the voltage times the amperage.

$$\text{Power} = \text{Voltage} \times \text{Amperage}$$

$$\text{Watts} = \text{Volts} \times \text{Amps}$$

$$W = V \times A$$

Using a household circuit of 20 amps and 120 volts, we can substitute and find the power.

$$W = 115V \times 20A$$

$$W = 115 \times 20$$

$$W = 2300$$

Since power is equal to current times voltage (Watt's Law), and current is equal to voltage divided by resistance (Ohm's Law), we can substitute the elements and state Watt's Law in terms of current and resistance:

$$W = I^2 \times R$$

We can also state the power law in terms of voltage and resistance:

$$W = \frac{E^2}{R}$$

$$\text{Amps (I)} = \frac{P}{E} \text{ or } \frac{E}{R} \text{ or } \sqrt{\frac{P}{R}}$$

$$\text{Volts (E)} = \sqrt{P \times R} \text{ or } I \times R \text{ or } \frac{P}{I}$$

$$\text{Ohms (R)} = \frac{E^2}{P} \text{ or } \frac{E}{I} \text{ or } \frac{P}{I^2}$$

$$\text{Watts (P)} = E \times I \text{ or } I^2 \times R \text{ or } \frac{E^2}{R}$$

Algebraic Manipulation of Ohm's Law
and the Power Law
Table 2-2

Additional Formulas

Further algebraic manipulation of Ohm's Law and the Power Law produces additional formulas. Table 2-2 gives a summary. The formula to use for a particular calculation depends on the values you have and what you need to find.

Voltage Drop

Voltage drop is a reduction in voltage between the power source and the load. This loss occurs whenever current flows in wires. It is equal to the product of the current and the resistance of the wires.

Resistance in wires is determined by their material, diameter and length. (See Table 2-3). For example, No. 12 copper wire has a resistance of 0.0016 ohms per foot (rounded up from 0.00159). If 20 amps flow through 400 feet of this wire, the

Size AWG Or MCM	Circular Mills	Copper		Aluminum	
		Weight per 1000 Ft.	Resistance ohms Per 1000 Ft.	Weight Per 1000 Ft.	Resistance ohms Per 1000 Ft.
AWG					
12	6530	19.8	1.59	6.01	3.67
10	10380	31.43	0.9988	9.556	1.62
8	16510	49.98	0.6281	15.20	1.02
6	26240	79.44	0.4952	24.15	0.6395
4	41740	126.3	0.2485	38.41	0.4021
3	52720	159.3	0.1971	48.43	0.3189
AWG					
2	66360	205.	0.1594	62.3	0.2580
1	83690	259.	0.2365	78.6	0.2045
0	105600	326.	0.1002	99.1	0.1622
00	133100	411.	0.07949	125.	0.2386
000	167800	518.	0.06306	157.	0.1020
0000	211600	653.	0.05999	199.	0.0809
MCM					
250	250000	772.	0.04232	235.	0.06847
300	300000	925.	0.03526	282.	0.05706
350	350000	1080.	0.03022	328.	0.04891
400	400000	1236.	0.03645	375.	0.04280
500	500000	1542.	0.02115	469.	0.03423
600	600000	1850.	0.01764	563.	0.02853
MCM					
700	700000	2160.	0.01512	657.	0.02445
750	750000	2316.	0.01411	704.	0.02283
800	800000	2469.	0.01322	751.	0.02140
900	900000	2780.	0.01175	845.	0.01902
1000	1000000	3086.	0.01058	938.	0.01712

Sizes, Weights and Resistances of Conductors at 20° C./68 Degrees F.
Table 2-3

voltage drop would be: ·

0.0016 ohms x 400 ft. x 20 A = 12.9 V

If 230 volts are available at the source of this wire, there will be 230 minus 12.9 volts, or 217.1 V at the point of use 400 feet away. NEC regulations specify permissible voltage drop in various circuits. Try to keep over-all voltage drop at 5% or less. This can be divided into 3% for feeders and services and 2% for branch circuits. See Table 6-24 in Chapter 6 for voltage drop calculation.

Series Circuits

In a *series circuit* the loads are connected *in the line* (See Figure 2-4). The loads can be thought of as links in a chain. If any link breaks, the chain breaks. For example, if three incandescent lamps are connected in a series circuit and one burns out, the current in the circuit is halted. The other lamps, though not defective, will also stop burning.

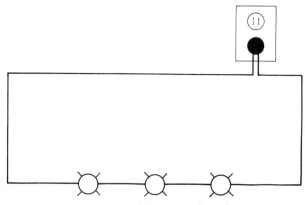

Three Lamps in a Series Circuit

Figure 2-4

Series circuits aren't used for most modern lighting and power loads. However, series calculations are necessary for a few specialized circuits and in the design of feeder circuits. Also, note that in every circuit the circuit wires themselves have resistance in series along with the load resistances.

In calculating series circuits, the following rules apply:

• The same current flows in all parts of the circuit.

• The voltage across each load adds up to the total voltage.

• The effective load-resistance of the circuit is equal to the sum of the individual loads.

Ohm's Law can be modified slightly to get the following formula:

$$I = \frac{E}{R_1 + R_2 + R_3 + R_4 + R_5 + etc.}$$

I = current flowing in all parts of the circuit

E = the total voltage found in the circuit

R_1, R_2, R_3, etc. = the individual resistances in the circuit

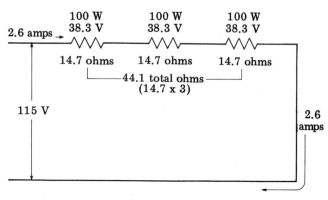

Schematic of Three Lamps in a
Series Circuit—Example 1

Figure 2-5

Example 1 (Figure 2-5): A 115-volt circuit with three, 100-watt lamps connected in a series. What is the total resistance? What are the individual resistances? What is the current?

Since the known factors here are volts and watts, use one of the power formulas to find the total resistance:

$$W = \frac{E^2}{R} \quad \text{is suitable.}$$

W	=	100 x 3 =	300 watts total
V	=	115	

$$300 = \frac{115 \times 115}{R}$$

$$300 \times R = 13225$$

$$R = \frac{13225}{300}$$

$$R = 44.1 \text{ ohms}$$

The value 44.1 ohms is the total resistance. In a series circuit the total resistance is the sum of the individual resistances. Therefore, to find individual resistance for each lamp, divide the total resistance by three:

$$\frac{44.1}{3} = 14.7 \text{ ohms per lamp}$$

Also, in a series circuit the same current flows in all parts of the circuit. Use Ohm's Law to find the current in Example 1, now that the total resistance is known.

$$I = \frac{E}{R}$$

$$I = \frac{115}{44.1}$$

$$I = 2.6 \text{ amps throughout the circuit}$$

In the example, the resistance of the wires was disregarded. The voltage drop occurring within the wires was also ignored because the distance was small. However, the reduction across each load in a series circuit can also be called voltage drop. Since the voltage across each load must add up to the total voltage, we can calculate this voltage drop easily:

$$\frac{115}{3} = 38.3 \text{ V per bulb}$$

Parallel Circuits

In a *parallel circuit* the loads are connected *across the line*. (See Figure 2-6) The failure of one load won't affect the others in the circuit.

Almost all modern lighting and power loads have parallel circuits. Parallel circuits differ from series circuits in several important ways. The following rules apply for parallel circuits:

• The same voltage flows in all parts of the circuit (small voltage drop within the conductors disregarded).

• The current across each load adds up to the total current.

• The effective total resistance of a group of resistances will always have a value less than that of the smallest resistance of the group.

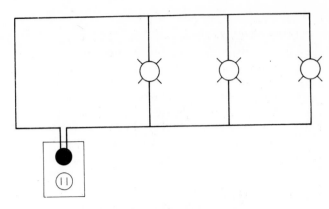

Three Lamps in a Parallel Circuit

Figure 2-6

Total effective resistance for a parallel group of resistances can be found from the following formula:

$$\frac{1}{R_{eff.}} = \frac{1}{R_1} + \frac{1}{R_2} + \frac{1}{R_3} \ldots \text{ etc.,}$$

where $R_{eff.}$ = the total effective resistance and R_1, R_2, R_3, etc. = the individual resistances.

It may be useful to think of the total effective resistance as an imaginary load that allows the same current to flow as the individual connected loads combined.

Calculations for parallel circuits usually involve using resistance and voltage to determine the ampacity of the circuit. Use Ohm's Law and related formulas to determine current and resistance values for the entire circuit or for individual loads.

Since most individual loads are given in watts, the following is a helpful shortcut:

substitute $\dfrac{E^2}{P}$ for R.

$$\frac{1}{R_{eff.}} = \frac{1}{R_1} + \frac{1}{R_2} + \frac{1}{R_3} \ldots \text{etc.}$$

$$\frac{1}{R_{eff.}} = \frac{1}{\dfrac{E_1^2}{P_1}} + \frac{1}{\dfrac{E_2^2}{P_2}} + \frac{1}{\dfrac{E_3^2}{P_3}} + \ldots \text{etc.}$$

$$\frac{1}{R_{eff.}} = \frac{P_1}{E_1^2} + \frac{P_2}{E_2^2} + \frac{P_3}{E_3^2} + \ldots \text{etc.}$$

The resistance of any electrical device is fixed by its design. You can find the device's design voltage on the nameplate. Therefore, the values for E in the above formulas could be different if each of the connected loads was a device with a different design voltage. However, manufacturing tolerances of 5% are usually allowed between design and circuit voltages. If the differences between the design and circuit voltages are small, the circuit voltage can be used. If this is the case, or if all loads have the same design voltage, the formula can be simplified:

$$\frac{1}{R_{eff.}} = \frac{P_1 + P_2 + P_3}{E^2} + \ldots \text{etc.}$$

where P_1, P_2, etc., are the rated wattages and E^2 is the design or circuit voltage.

Example 2 (Figure 2-7): What size appliance circuit is needed to supply the given connected load?
A toaster oven - 1350 watts
A coffeemaker - 600 watts
A blender - 250 watts

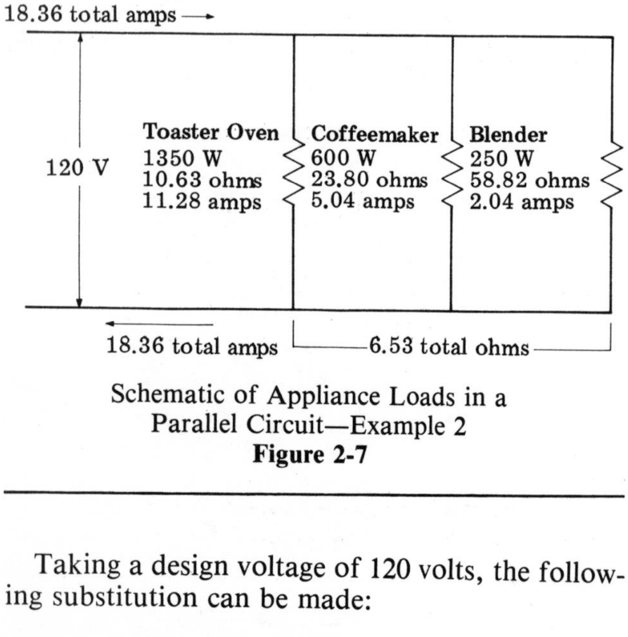

18.36 total amps →

	Toaster Oven	Coffeemaker	Blender
120 V	1350 W	600 W	250 W
	10.63 ohms	23.80 ohms	58.82 ohms
	11.28 amps	5.04 amps	2.04 amps

18.36 total amps ——————— 6.53 total ohms

Schematic of Appliance Loads in a
Parallel Circuit—Example 2
Figure 2-7

Taking a design voltage of 120 volts, the following substitution can be made:

$$\frac{1}{R_{eff.}} = \frac{\text{total rated wattage}}{(\text{design voltage})^2}$$

$$\frac{1}{R_{eff.}} = \frac{P_1 + P_2 + P_3}{E^2}$$

$$\frac{1}{R_{eff.}} = \frac{1350 + 600 + 250}{(120)^2} = \frac{2200}{14400}$$

$$\frac{1}{R_{eff.}} = .153 \text{ ohms}$$

Using Ohm's law:

$$I = \frac{E}{R_{eff.}} = E \times \frac{1}{R_{eff.}}$$

$$I = 120 \times .153 = 18.36 \text{ amps}$$

A single 20-amp appliance circuit would be large enough to supply these appliances.

Occasionally you'll need to find current and resistance values for individual loads. These can be calculated by substituting the desired single load wattage in the above steps.

DC and AC Current

DC, or *Direct Current*, is the flow of current through the conductor at a constant rate, in one direction. A dry cell battery connected to a light bulb is a simple DC circuit.

AC, or *Alternating Current*, is the flow of current through the conductor first in one direction, then in reverse. The *frequency*, or rate of reversal and flow, depends on the voltage source. The unit of measure of this frequency is called a *Hertz* (Hz).

Almost all power systems in the United States operate at a frequency of *60 cycles per second*. One cycle is made up of two half-cycles. Each half-cycle represents the flow of current in one direction.

To reverse itself, each half-cycle must begin and end. It doesn't do this at once. Rather, it starts at zero, accelerates to a peak, then slows again to zero. A good comparison, although much slower, is like driving a car from a stop to a certain speed and then slowing down to a stop again. A convenient way of diagraming this process is to use a *sine wave* diagram, as shown in Figure 2-8.

In this diagram one complete cycle is shown. A half-cycle contains 180 degrees, the complete cycle 360. The rate of each half-cycle is shown beginning at zero at the axis, peaking at 100% at 90 degrees, and then returning to zero. The first half-cycle is given a positive value, the second a negative one.

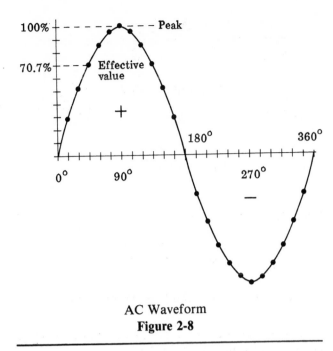

AC Waveform
Figure 2-8

The value of current in an alternating current wave is constantly changing. The only way to evaluate it is to find its *effective* value. This is done by comparing the heating effect of AC current with the heating effect of DC current. A common description of effective value is the *rms* value. The letters "rms" stand for "root-mean-square," and come from the math formula used to compare AC and DC heating effects.

Since the cycles in AC current occur very fast and at a regular rate, the effective value of the current can be treated as if it were a steady flow of current. Ohm's Law and related formulas will apply. However, since it represents a changing rate, an effective or rms value does not represent a maximum current flow. The effective value of a sine wave alternating current is equal to 0.707 times the maximum, or peak value, of the current wave.

When an AC current is given a value, it will always be the effective value unless otherwise noted. Reference tables and books, equipment nameplates, test instruments and current ratings of circuits and equipment are in effective values.

Power Factor

Current by itself is meaningless. Unless it's put to work it has no value. To be put to work, it must be pushed through the circuit by the voltage.

The flow of voltage through an AC circuit also occurs in cycles. This can be shown with a sine curve. When its path is similar to the amperage—that is, when highs and lows of the amperage and voltage curves take place at the same time—the circuit is said to be "in phase." (See Figure 2-9.)

When a circuit is "in phase," the power that results is considered *real*, or *positive*, power. In a circuit with real power, the value shown on a wattmeter would be the volts times the amps. The test reading would show the power available in the circuit.

Power also has a negative value. Unlike current, the positive and negative values of power *do not* result from the frequency or cycles in the circuit. In a circuit "in phase," power produced by both the positive and negative halves of the cycle, or curve, is considered real power. The negative and positive values of power are determined by the relationship between voltage and amperage *within* the cycle. This relationship is affected by the types of loads on the circuit.

Three types of loads can be applied on a circuit: resistive, inductive and capacitive. Light bulbs and toasters are examples of resistive loads. Resistive loads will always have an "in phase" circuit. Inductance and capacitance are found in motor loads as a result of electromotive forces. Inductance and capacitance differ from resistive loads in several ways. One is that they can put the circuit "out-of-phase." When this happens, the current and voltage travel at different rates. The current is said to either *lead* the voltage or *lag* behind it.

If the current either leads or lags the voltage by 90 degrees, the result is a cycle in which the power is half positive and half negative. These values cancel each other and the circuit is essentially powerless.

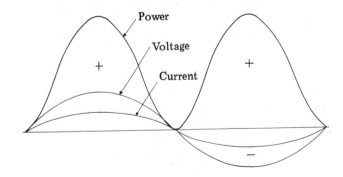

Current, Voltage and Power "In Phase"
Figure 2-9

Although the average power in such a circuit is zero, there is a measurable power surging aimlessly through the circuit. This is *reactive power* and can be measured by a *volt-ampere-reactive*, or *VAR meter*. The unit value is also called a VAR. One VAR is numerically equal to one watt. But a VAR can't do any work.

You'll seldom find a circuit that's entirely resistive or inductive-capacitive. Most circuits are combinations of the two. As a result, the power within a circuit is a combination of real and reactive power. See Figure 2-10. This combination is called the *volt-ampere*, or *apparent*, power. To find its value, multiply the volts times the amps, just as for real power in an in-phase circuit. It's important to note, however, that the circuit is no longer purely resistive. It is no longer "in phase." If a wattmeter is applied, the meter will read only the real power; it will not read the reactive power. The value shown will be less than the apparent or calculated power.

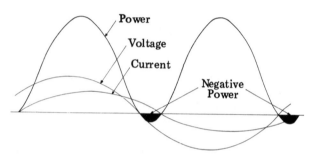

Positive and Negative Power in an
"Out-Of-Phase" Circuit
Figure 2-10

This relationship between real power and apparent power is called the *power factor* of the circuit. Its relationship can be expressed numerically by the following formula:

$$\text{Power Factor (PF)} = \frac{\text{Watts of Real Power}}{\text{Volts-Ampere}}$$

For example, if an electrician uses a volt-amp-wattmeter and discovers the load in a circuit is drawing 8 amps at 115 volts, with a real power reading of 500 watts:

$$\text{PF} = \frac{500}{8 \times 115} = \frac{500}{920} = 54\% \text{ approximately}$$

Power factors are useful in determining the efficiency of an electrical system. When a system has a power factor of 100%, it's operating at peak effi-ciency. In some areas, utilities charge a penalty to customers whose systems use induction motors with low power factors. These systems can be corrected with *capacitors*. A capacitor produces a leading current which cancels the reactive part of the current drawn by other loads in the circuit. This permits the line current to act as if only resistive loads were in the circuit.

Grounding
Grounding provides a way of conducting electrical current in a system to the earth. Without the ground, leaking current will try to find some other conducting body to travel through. If this body happens to be human, dangerous and often fatal shocks can occur. Don't underestimate the importance of a ground wire.

A ground is, at its simplest, another conductor in a circuit. But while "hot" wires in a circuit carry the current to where it can do useful work, the ground wire does the opposite. It provides a path of "no-work," or a path of lowest practical resistance. This low resistance is called the *ground potential* of the circuit.

In addition to minimizing shocks, grounds provide an alternate route to the panel housing fuses or breakers. This alternate route is important if there is a transformer malfunction, if lightning strikes the transmission wires, or if there is a fault in the system itself.

Under normal circumstances, ground wires don't carry current. They're often called "neutral wires" and are color-coded white or green.

Over the years the NEC has upgraded requirements for grounding. Today all electrical systems must be properly grounded.

Short Circuits and Ground Faults
Electrical faults are common when the insulation around a conductor breaks down, either through abuse or aging. When this happens, a *short circuit* results. A short can also result from improperly connected wires. Recently the term *ground fault* has come into use, to distinguish it from the faults occurring within a circuit. Technically, a short circuit happens when the conductors touch. A ground fault, on the other hand, happens when the conductor touches its enclosure. A wire shorting to its conduit or to the metal housing of an appliance is a ground fault. Two wires touching in a worn appliance cord is a short circuit.

A *ground fault circuit interrupter* is now required for areas where dangerous electric shock is likely to occur because of dampness, such as outdoor and bathroom receptacles. It's a special circuit breaker, installed either at the panel or at the receptacle itself, which detects a ground fault and opens the circuit. See Figure 2-11.

In a typical circuit, the breaker or fuse doesn't open until the current exceeds 15 or 20 amps. But when a ground fault occurs, the line current may not be excessive. The breaker or fuse may not trip. However, a person can suffer a fatal shock from an ampere flow as small as 1/15 of an amp. The GFCI is designed to open when there is a very small ground fault current flow. It also opens very quickly, in about 1/8 of a second. That's quick enough to prevent what might otherwise be an accidental electrocution.

Courtesy: Leviton Manufacturing Co.

15 amp, 120 volt GFCI Receptacle
Figure 2-11

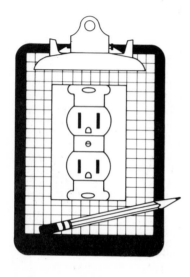

Chapter 3:

Tools and Testing Instruments

Every electrician's tool kit includes several hand tools used by all building trades. It also has tools designed specifically for electrical work. A basic set of tools will handle most electrical work. More specialized jobs require more specialized tools.

Many of the tools electricians use are not essential, they just make the work easier and faster. Some equipment also falls into this category. For example, power conduit benders and rolling scaffolding speed the work and increase productivity. But they're expensive, may not be used more than a few times a month, and take up valuable storage space both in your shop and on the site. It may be better to rent rather than buy this equipment.

My advice would be to buy slightly less than all of the tools you think you need, but be sure what you buy is good quality.

Keep your tools in good condition. Chisels, saws and knives should be sharp. Screwdrivers are a primary electrical tool. Their blades should be sharp and not bent or twisted. Electrical testing equipment must be in good working order. Defective instruments will not only lead to false readings, they can cause severe shock.

Tools and equipment should be kept clean and rust-free. A neat, orderly toolbox protects tools and makes the work go faster.

Drills and Bits

A 3/8- or 1/2-inch electric drill is part of every electrician's tool kit. This size will handle most work. For heavy-duty drilling, a larger drill may be needed.

Another handy tool is a *brace*. This permits drilling through wood when power is not available.

You should have a set of drill bits ranging from 1/2- to 1-inch for most hole-cutting in joists and panels. Standard sizes have a 6'' to 8'' shaft. Longer sizes may be need for special jobs. For sheet metal and steel, you'll need a complete set of carbide steel bits. Mounting panels and fasteners onto block or concrete walls will require different-size masonry bits.

A 100-foot grounded extension cord saves time because it allows a large amount of drilling from a single outlet.

Hammers

Hammers are classified by the weight of the head and by the type. Have at least one 16-ounce claw hammer in your tool kit for driving staples and nailing hangers. If the job requires chiseling openings in concrete block, use a heavier "blacksmith" hammer. Always wear safety glasses when hammering or chiseling against metal or masonry.

Chisels

Cold chisels are used on metal and steel. 1/2-inch and 3/4-inch chisels are about right for most jobs. Keep your chisels sharp, and trim them occasionally on a grinder to prevent "mushrooming" of the heads.

You'll also need a 3/4-inch wood chisel to notch joists and braces for cable or conduit. And you should have several masonry chisels if you do a lot of work on concrete or block.

A center punch, used to score metal and steel to position a drill bit, should also be on hand.

Other sizes and kinds of chisels can be added as needed.

Saws

A hacksaw is needed to cut conduit or cable. Choose one with a large, sturdy grip. Extra blades are essential.

A keyhole saw has a thin, pointed blade. It's useful for cutting paneling or drywall to shape openings for outlets and junction boxes. For larger, more difficult jobs, use a carpenter's crosscut saw. Be sure to keep all saws sharp and free from rust.

Wrenches

A 10-inch adjustable wrench, commonly called a crescent wrench, fits most nuts and bolts. Larger and smaller sizes are available. Vise-grip pliers are useful for firmly gripping pipe or other materials. They can also hold materials together, freeing your hands for other work. Vise grips come in many sizes. If you work with threaded conduit, you may need several sizes of pipe wrenches.

Wrenches should be cleaned and oiled occasionally. Avoid hammering on them to free a stubborn nut or bolt. When working around live equipment, be careful that the wrench doesn't slip and put you in contact with a hot wire.

Screwdrivers

Screwdrivers will be the most-used tool in your pouch. They should be chosen with care. Several blade widths, thicknesses and lengths are necessary. The grip is also important. Plastic grips don't break as easily as wood. But some electricians feel a wooden grip is easier to hold, especially when your hands are moist. Whatever grip you choose, make sure the diameter is large enough. A grooved grip gives a better hold.

Avoid screwdrivers with a blade shaft exposed at the top of the grip. Plastic and wooden grips with concealed shafts provide protection from shock. Some grips have a rubber covering for extra protection.

A ratcheting screwdriver makes screwdriving quicker and easier.

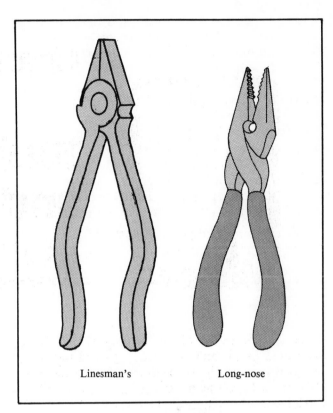

Types of Pliers
Figure 3-1

Pliers

The two basic pliers needed for electrical work are the blunt-nosed linesman's pliers and the smaller long-nosed pliers. See Figure 3-1. Be sure the grips fit your hand comfortably. Check the serrations on the tips of the jaws. They should be deep and not too fine.

These pliers often have blades on the jaws for cutting wire. Some electricians, however, prefer wire clippers or cutters for this.

Most electrician's pliers have insulated handles. This helps protect against shock. But don't assume that an insulated handle will always protect you against any jolt. Other ways to protect against shock are detailed later in this chapter.

Slip-joint pliers are handy for nuts and bolts. A drop of oil on the joints of all pliers keeps them opening easily.

Rules and Measuring Tapes

A six-foot wooden folding rule is handy for measuring wire, conduit and openings. It's rigid, so one man can measure most dimensions up to 6 feet. For longer lengths, a 25-foot steel measuring tape is needed.

Keep tapes and rules clean and dry. An occasional drop of oil on a steel tape increases its life. Be sure to buy a steel tape which has replacements for the tape itself.

Knives and Wire Strippers

A sharp jackknife is a must for an electrician. It's used for trimming wire insulation, cutting tape and trimming openings. Some electricians carry a utility knife. These knives have blades which draw back into the handle. The blades are thin and very sharp but not as strong as a jackknife. However, replacement blades are inexpensive and can be carried in the handle. Dull and broken blades are easily replaced.

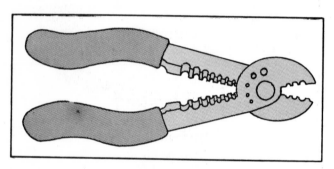

Combination Wire Stripper and Cutters
Figure 3-2

A useful tool when working with wire is the combination cutter and stripper (Figure 3-2). This tool, shaped something like pliers, has various-size openings that strip wire clean of insulation. The openings are labeled for various wire sizes. It also has jaws to crimp solderless terminals.

Tape

Many electrical connections require taping. Taping used to involve layers of rubber tape covered with layers of friction tape. Most taping now is done with plastic tape. This tape provides insulation, weatherproofing and durability. Rolls are available in plastic dispensers, making the tape easy to peel off and cut.

Fuse Puller

Fuse pullers reduce the danger of shock when replacing cartridge fuses. See Figure 3-3. They are also used to handle live electrical parts. Fuse pullers are made of plastic or a fiber material.

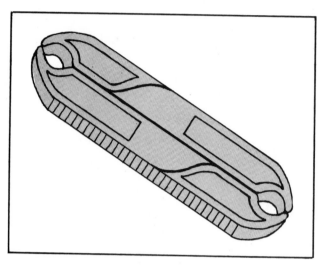

Cartridge Fuse Puller
Figure 3-3

Files

Small and medium mill files, rounded and flat, remove burrs from conduit and other metal equipment. A burr or rough edge can damage the insulation when wire is pulled or installed, resulting in a ground fault.

Files can also shape and sharpen other hand tools and equipment on the job.

Tool Aprons and Belts

Many electricians use a tool apron to carry hand tools. Some prefer not to, complaining about weight and awkwardness. An apron should have enough pockets and slots to carry the hand tools you use most. Most aprons are leather. Be sure to keep them dry and oiled. With proper care, they'll last many years.

The belt which carries the apron should be wide enough to support the weight with comfort. Its clasp should open and close easily.

Conduit Tools

There are several types of electrical conduit. Flexible metallic conduit is cut with a hacksaw. Rigid steel conduit (RSC) can also be cut with a hacksaw, but it's neater and easier to use a pipe cutter. Electrical metallic tubing (called EMT or thin-wall) should be cut with a thin-wall conduit cutter. This prevents pinching the tube. After the conduit is

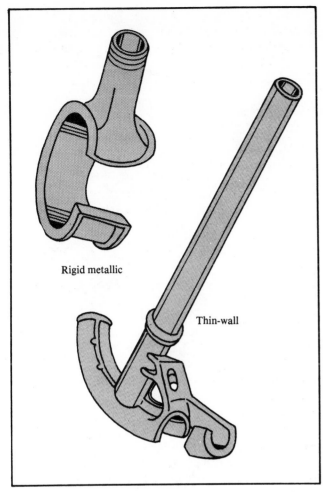

Conduit Benders
Figure 3-4

Rigid non-metallic conduit is now used in many installations. It's made of PVC or fiberglass and can be cut easily with a hacksaw. Threaded couplings are cemented to the conduit. They hold it to outlets and boxes. PVC conduit can be bent on site with a special portable oven. It bends very easily into the desired shape. Common pre-bent shapes can also be purchased.

To bend rigid metallic and thin-wall conduit, you'll need a conduit bender. See Figure 3-4. This tool is also called a *hickey*. It bends the conduit without kinking it. The hickey is operated by hand and is used on most small diameter conduit. For larger diameters, you'll probably need an electrically driven bender.

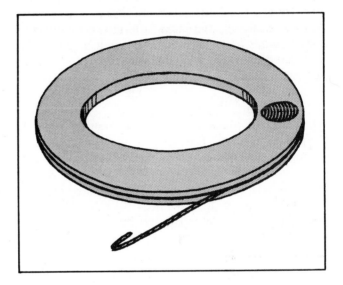

Fish Wire
Figure 3-5

cut, use a reamer or file to remove burrs inside the openings.

Rigid metallic conduit is threaded for installation. This requires taps and dies and a vise to hold it securely.

Most thin-wall conduit isn't threaded. A tool called an *indenter* or *impinger* attaches couplings and other fittings. This tool, when pressed against the fitting, forms indentations in the fitting and the wall of the tubing. Compression-type couplings are also used. These come in cast metal and steel. A threaded bushing forces a sleeve against the tubing and the inside of the coupling to hold it in place. A third type of thin-wall coupling uses a machine screw.

Take care to install thin-wall couplings properly. Be sure compression nuts and screws are tight, giving a low-resistance connection between the coupling and the conduit. Chapter 7 gives tips on proper conduit installation.

Fish Wires and Drop Chains
A *fish wire* is used to pull wire through conduit. It's made of tempered spring steel about 1/8 inch wide and is available in various lengths. The fish wire is pushed or pulled through the conduit. The electrical wire or wires are then tied to a hook on the end. The "fish" is pulled back through, bringing the wires along with it. A reel-type holder keeps the fish coiled when not in use. See Figure 3-5. The fish wire is also used in existing buildings to pull cable between floors and in walls.

A *drop chain* is a small-link chain with a lead or iron weight. Although not used as much as the fish wire, it's handy for feeding wire through vertical openings. It's also used to locate horizontal braces or obstructions in walls.

Testing Instruments

You have a wide choice of testing instruments for many types of applications. A few basic instruments will work for most jobs. But instruments with special features are available if you need them.

Choose quality instruments. They will last longer and need fewer repairs. And using them will be easier and safer. The cases should be sturdy. Leads, jaws and probes should be heavy enough and well insulated. Numbers and scales should be large enough to read easily.

Treat the instrument with care. Rough handling will cause breakdowns. Keep probes and clips clean, dry and free of oil. Replace defective clips and probes at once. If the instrument uses a battery, check it occasionally to make sure there's enough charge. Keep instruments and their accessories in their cases when not in use.

Many tests must be made on live circuits. Safety is a primary consideration. Read the instrument's instruction book carefully. This will help ensure safe operation. Become familiar with all switches and controls. Make sure the instrument is working properly by testing it on a *known* live line. You can be fooled into thinking a live line is dead if the instrument isn't working properly.

Shock is caused by current passing through you to ground. Caution and common sense are the best ways to prevent shock. When hooking up an instrument, be careful where you stand. A wet floor or wet shoes can result in severe shock if you accidentally touch a terminal. Water on a roof can be hazardous when testing rooftop equipment. Drain pipes, conduit and other metal parts allow a nice path to ground.

You can interrupt this path by using insulating materials, such as a wooden stepladder or wooden crates and pallets to stand on. Be sure they're dry and that they don't have metal bolts or reinforcing bands extending through them. Some electricians carry pieces of foam-rubber-backed carpeting to cover steel framework or damp concrete.

These precautions alone won't provide complete protection. But they'll help reduce the danger if you contact a live wire or terminal.

An instrument with clip-type leads requires that your fingertips be close to the terminal when making the connection. Since the clips will remain in place once attached, turn off the power when connecting them. Then switch the power on to get the reading. If this isn't possible, wear insulated gloves. Don't force the clips. When applying force, your fingers are likely to slip. If the current must be on, look for an easier connection.

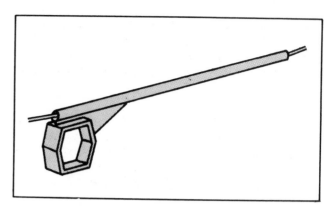

Insulated Instrument Probe
Figure 3-6

When using probes, apply light to medium pressure in a straight line. Try not to probe past other connections. Find a different connection instead. Many electrical suppliers stock insulated probe handles which have loops for your fingers. See Figure 3-6. These prevent slipping and keep your hands away from hot connections. Their small cost is money well spent.

A broken wire inside the instrument lead can give a false reading. You can be fooled into thinking a hot line is safe to work on. To check the leads, take a reading from a known hot connection to ground, with the lead pulled taut. If the needle doesn't move, replace the lead with a new one. If the needle still doesn't move, the instrument may be defective.

Keep spare leads, probes, clips and batteries in your toolbox. Replace broken parts at once.

When using an instrument to make resistance checks, make sure the power in the circuit is off. Checking resistance in a live line will blow the fuse in the instrument. If this happens, replace the fuse with one of the proper size. Using a fuse of the wrong size can cause serious damage to the instrument.

Many instruments for measuring amperage have locking jaws which snap around a conductor. This usually presents no problem, unless there's lack of space in the panel or box. If this is the case, the first impulse is to pull on the wire. But before doing so, throw the switch or breaker. Then pull on the wire carefully to prevent bending or breaking the terminals.

Instrument jaws should not be clamped to a metal box or used to support the instrument. They're not intended for this, and damage can result. Current can leak through the jaws into the metal box, causing a dangerous situation.

Some instrument checks are made over a period of time. If any check requires leaving a panel or box open, label it "Danger - high voltage" to inform anyone who may not be aware of the danger.

Test Lamp

A test lamp is the simplest of instruments. It consists of two probes and a lamp which glows when voltage is in the circuit. It's used to find live circuits and connections, to locate shorts in wires and connections, to test fuses and to locate the ground lead.

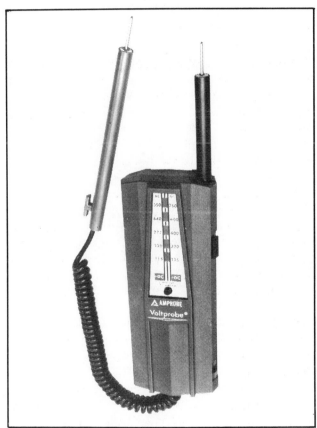

Courtesy: Amprobe Instrument Co.

Voltage Tester
Figure 3-7

Voltage Testers

A voltage tester is like a test lamp but more precise. It has a calibrated scale that indicates the voltage flowing through a circuit. See Figure 3-7. Most models are adjustable to handle both AC and DC circuits. Different models cover a wide range of voltages.

To be used properly, a voltmeter must be connected in parallel—across the load. See Figure 3-8.

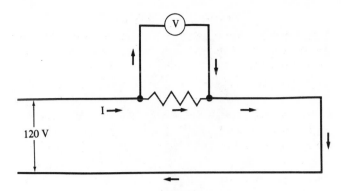

Voltmeter Connected Across the Load
Figure 3-8

The circuit must be closed ("hot") to allow the voltage to flow. Connections for voltage are made through probes or snap-on clips.

Ammeters

Ammeters measure the current flowing in a circuit. An ammeter must be connected in series in the circuit. See Figure 3-9. Connecting it across the load will destroy the instrument. Current must flow through the circuit to get a reading.

Many ammeters have jaws which open to snap around the conductor. Note Figure 3-10. This is the only connection necessary. The jaws act as transformers and give the proper reading on the meter. Clamp-on ammeters can be used only on AC circuits. They can measure only one conductor at a time.

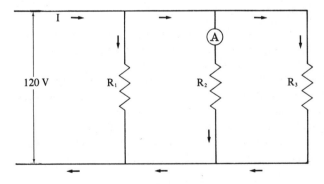

Ammeter Connected In the Line
Figure 3-9

Ohmmeters

An ohmmeter measures the resistance across a load. They are never used on a live circuit. Connecting an ohmmeter across a live voltage source will

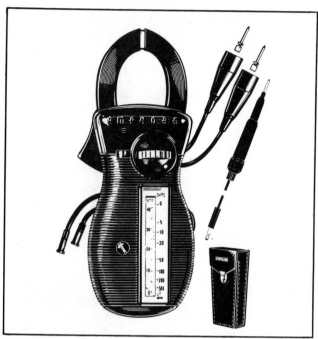

Clamp-On Ammeter
Figure 3-10

either destroy the instrument or blow its fuse. Ohmmeters are battery powered and use probes or clips as connectors.

Ohmmeters measure the resistance in a conductor or device. Since resistances can be calculated or are given on the design nameplate of the device, ohmmeters are useful in tracking down broken

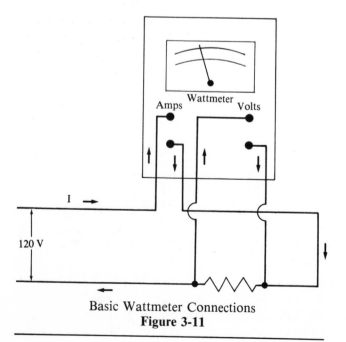

Basic Wattmeter Connections
Figure 3-11

wires or motor windings, defective coils and other items which may overload a circuit.

Wattmeters

Wattmeters measure power in a circuit. Since power is a product of current and voltage, most wattmeters have scales to indicate amps and volts. These instruments are called volt-amp-wattmeters. The voltage leads are connected across the load. The current leads are connected in series. Power must flow in the circuit. See Figure 3-11.

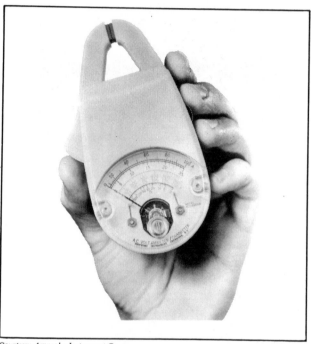

Pocket-Sized Multimeter
Figure 3-12

Many wattmeters have a special adapter cord and plug. If the load to be tested is a plug-in appliance or device, it's unplugged from the receptacle and plugged into the wattmeter. The wattmeter is then plugged into the receptacle. Power drawn by the appliance can be measured without disconnecting any wires or making any additional connections.

Multimeters

Multimeters combine the functions of two or three instruments. A volt-ohm-ammeter is a good example. See Figure 3-12.

Simple volt-ohm-ammeters measure the AC values in residential and light commercial systems. Most complex multimeters will measure AC and

Courtesy: Amprobe Instrument Co.

An Industrial Multimeter
Figure 3-13

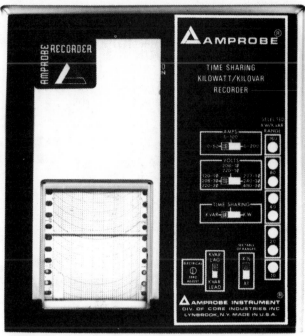

Courtesy: Amprobe Instrument Co.

Kilowatt/VAR Meter With a
Strip-Chart Recorder
Figure 3-14

DC voltages, ohms, temperatures and AC and DC current. Special accessories will permit kilovolt AC and DC ranges. Resistance ranges on multimeters can run as high as 5 megaohms (50,000 ohms). See Figure 3-13.

Connections on multimeters are made with jaws, probes, or clips. They follow the same rules as the more basic instruments. Current connections are made in series; voltage connections in parallel.

Multimeters have range switches. The switch should always be set on a range equal to or larger than the value to be tested. If the value is unknown or in doubt, the switch should be set to the highest range. Then set the range lower until the correct reading is taken. This avoids damaging the instrument.

Many multimeters now have digital read-out scales. Some have strip-chart recorders as in Figure 3-14. These recorders provide a written record of readings over a period of time.

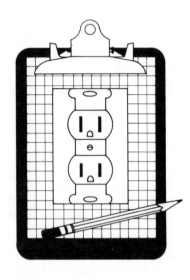

Chapter 4:

Materials

There's a lot to know about electrical materials. Fortunately, you don't need to remember it all. Knowing where to find the information is nearly as good as knowing the answer most of the time. Let this chapter be your handy guide to the most common electrical materials.

The most basic material for all electrical work is wire and cable. So that's where we'll start.

Wire

Wire is the basic conductor of electricity. *Solid wire* consists of a single length of drawn metal. *Stranded wire* is made up of solid wire twisted or braided together.

Electricians often refer to a large-diameter single or stranded wire as *cable*. The word cable, however, more correctly describes two or more individually insulated wires twisted or running together. Often the wires are enclosed in an additional plastic or rubber covering. Cable is explained later in this chapter.

Copper vs. Aluminum

Copper is the most common metal used for electrical wire. It's also the most practical when cost, conductivity and strength are considered.

But aluminum wire has advantages. It's lighter and less expensive than copper, but it's not as good a conductor. Some electricians feel that it's not as

strong as copper and complain that it breaks easily. Also, if not installed correctly, aluminum wire may overheat.

The debate over use of aluminum or copper wire may never be settled. It's important for you to note that there are two conductors with similar methods of installation but different properties and characteristics. You can then decide what's best for the job. Also check local codes.

First, thermal expansion of aluminum is 40% greater than copper. Heating a copper-aluminum joint puts pressure on the aluminum conductor. Second, aluminum is softer than copper and will deform under less pressure. This produces "cold flow" of an aluminum conductor in a heated copper-aluminum contact. When the electrical load is reduced and the contact cools, the lower contact pressure produces a high-resistance joint. This joint will tend to overheat under new loads. Third, bare aluminum surfaces exposed to air will oxidize. This forms a high-resistance film on the conductor which increases resistance and may cause overheating.

Aluminum wire is generally acceptable if you take the following precautions:

• Use only UL listed devices designed for use with aluminum wire. These are made of proper materials, with large bearing surfaces to reduce contact pressure.

• Use only new alloys of aluminum wire. These are identified on the box as "Aluminum Conductor Material" or "New Conductor Material."

• Select outlet boxes of the proper size for the number of wires to be installed (NEC Article 370).

• Use a wire stripper to remove the insulation. "Ringing" or notching that may result from using a knife can cause the wire to fatigue and fail.

• Do not use "push-in" devices that rely on spring pressure to hold the wire.

• Loop the wire 2/3 to 3/4 around the screw. Do it in the direction of the screw rotation. Tighten the screw until it's snug. Then add 1/2 turn—no more. *Do not overtighten.*

• Position wires carefully when fastening the device to reduce the loosening of the screws.

• Use only UL connectors stamped "CU/AL."

• Never use materials that haven't been approved for aluminum wire systems.

One type of wire is made of both copper and aluminum. It has a copper coating and an aluminum core. It's current-rated the same as aluminum wire and has an outer covering of copper equal to about 10% of the conductor cross-sectional area. The same rules that control installation of aluminum wire apply to copper-clad aluminum.

Ampacity
All wires have some internal resistance which causes power loss. Power is also lost as a result of heat. This is known as the I^2R *power loss.* Resistance and I^2R power loss depend on four items:
1. The type of wire—copper or aluminum
2. The cross-sectional area of the wire
3. The insulation around the wire
4. The location of the wire

These four items determine the ampacity of the wire—its ability to carry current.

Size and Gauges
The ability of wire to carry current is determined by its *cross-sectional area.* The cross-sectional area of the wire is found by squaring its diameter.

Most wires have small diameters, expressed in thousandths of an inch. 1/1,000 of an inch is a *mil.* The cross-sectional area of a wire is given in *circular mils.* One circular mil is the area of a circle one mil in diameter. See Table 4-1.

$$1 \text{ inch} = 1000 \text{ mils} \qquad \text{inches} = \text{mils} \times .001$$
$$1 \text{ mil} = .001 \text{ inch} \qquad \text{mils} = \text{inches} \times 1000$$
$$\text{Circular mil} = \text{diameter (in mils) squared}$$
$$cm = d^2$$

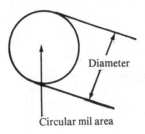

Circular Mil Area of Wire
Table 4-1

To convert the diameter of a wire from inches to mils, multiply the inch dimension by 1,000. For example, the diameter of a 1/8-inch wire would be:

$$1/8 \times 1,000 = \frac{1,000}{8} = 125 \text{ mils}$$

To find circular mils, square 125:

$$125^2 = 125 \times 125 = 15,625 \text{ cm}$$

Because such a large figure would be difficult to work with on the job, the electrical industry uses the American Wire Gauge system (AWG). It's also known as the Brown and Sharpe system (B&S).

Each circular mil wire size is given a number. Sizes begin at No. 40 and increase to No. 1. Sizes larger than No. 1 use the *ought* scale:

No. 0	—	One ought
No. 00	or 2/0 —	Two ought
No. 000	or 3/0 —	Three ought
No. 0000	or 4/0 —	Four ought

Four ought is the largest AWG size. Larger wires are given in circular mils. The abbreviation MCM is used to show 1,000 circular mils. For example, a wire of 300,000 circular mils is 300 MCM.

Most wiring systems use wires between No. 12 and 4/0. Also, it's the practice in electrical work to use only even numbered AWG sizes (Table 4-2).

The ability of wire to carry current increases as its size increases. Note that on the AWG scale, the larger numbers indicate smaller wires. Also, the larger the AWG number, the greater the resistance. A No. 12 wire, for example, is smaller than a No. 10. Its resistance is greater and it will carry less current. However, because it's smaller, it will cost less.

While a wire must be large enough to carry the load safely, oversizing can be costly. Also, the NEC regulates the number and size of wires running in any one raceway. (See the section on Conduit Restrictions in this chapter.) Oversize wires can increase the number of raceways needed and increase costs without adding any significant benefit.

Stranded Wire Gauges

Stranded wire is more flexible than solid wire. It's also easier to work with, which is why it's often used in installations requiring larger conductors. Ampacities of stranded wire are determined by taking the ampacity of the nearest equal cross-sectional area of solid wire. Because of this, a stranded wire with the same gauge as a solid wire will have a slightly larger diameter. In most cases this difference in size can be ignored.

Resistance

Most electricians aren't directly concerned with the resistance of a wire. They're concern is conductor resistance as it is related to ampacity and voltage drop. Resistance in ohms is given in AWG tables for occasions when the actual resistance of wire must be calculated.

Size AWG Or MCM	Circular Mills	Copper		Aluminum	
		Weight per 1000 Ft.	Resistance ohms Per 1000 Ft.	Weight Per 1000 Ft.	Resistance ohms Per 1000 Ft.
AWG					
12	6530	19.8	1.59	6.01	3.67
10	10380	31.43	0.9988	9.556	1.62
8	16510	49.98	0.6281	15.20	1.02
6	26240	79.44	0.4952	24.15	0.6395
4	41740	126.3	0.2485	38.41	0.4021
3	52720	159.3	0.1971	48.43	0.3189
AWG					
2	66360	205.	0.1594	62.3	0.2580
1	83690	259.	0.2365	78.6	0.2045
0	105600	326.	0.1002	99.1	0.1622
00	133100	411.	0.07949	125.	0.2386
000	167800	518.	0.06306	157.	0.1020
0000	211600	653.	0.05999	199.	0.0809
MCM					
250	250000	772.	0.04232	235.	0.06847
300	300000	925.	0.03526	282.	0.05706
350	350000	1080.	0.03022	328.	0.04891
400	400000	1236.	0.03645	375.	0.04280
500	500000	1542.	0.02115	469.	0.03423
600	600000	1850.	0.01764	563.	0.02853
MCM					
700	700000	2160.	0.01512	657.	0.02445
750	750000	2316.	0.01411	704.	0.02283
800	800000	2469.	0.01322	751.	0.02140
900	900000	2780.	0.01175	845.	0.01902
1000	1000000	3086.	0.01058	938.	0.01712

Sizes, Weights and Resistances of Conductors at 20° C./68 Degrees F.
Table 4-2

Resistance values in AWG tables are given for direct current. There's little difference between resistance values for DC and AC current in conductors No. 4/0 and smaller. When conductors larger than No. 4/0 are used in AC circuits, a multiplying factor must be used. These factors are found in the NEC (Chapter 9, Table 9). Since most residential and light commercial work uses No. 4/0 or smaller, the standard AWG table will work for most applications.

Temperature

Resistance in a conductor is related to its temperature. The resistance of most conductors increases as its temperature increases. Resistance values given in wire tables are usually at room temperature—20 to 25 degrees centigrade, which is equal to 68 to 77 degrees Fahrenheit. If a conductor is located in an area where the temperature is much higher or lower than normal, consider the effects this temperature may have on wire resistance.

The relationship between resistance and temperature is expressed by the formula:

$$R_x = R\frac{234.5 + t_x}{234.5 + t}$$

where R_x = the unknown resistance

R = the known resistance at t

t_x = the variable temperature in degrees centigrade

t = the known temperature in degrees centigrade

234.5 = math constant

For example, what is the resistance of No. 8 wire at 85 degrees Fahrenheit?

To use the above formula, convert degrees Fahrenheit into degrees centigrade:

Fahrenheit = 1.8 x centigrade + 32

85° = 1.8 x C + 32

85 - 32 = 1.8 x C

$\frac{53}{1.8}$ = C

C = 29.5°

Next the values are substituted into the formula:

R_x = the unknown resistance

R = .63 ohms per 1,000 feet (Table 4-2)

t_x = 85° F = 29.5° C

t = 20° C (Table 4-2)

$$R_x = .63\frac{234.5 + 29.5}{234.5 \times 20}$$

$$R_x = .63\frac{264}{254.5}$$

$$R_x = .63 \times 1.03$$

$$R_x = .653 \text{ ohms per 1,000 feet.}$$

Most electricians aren't directly concerned with wire temperature. However, if a wire is operating at a cooler temperature, due to either location or insulation, smaller conductors may sometimes be used.

Insulation

Wires are bare, covered or insulated. A bare wire is just a metal conductor. A covered wire has a plastic or weatherproof covering that doesn't have insulating abilities. Bare and covered conductors are generally found only in overhead, open-air wiring.

Insulated wire is required in nearly all residential and commercial systems. Insulated wire protects the area around it from heat caused by current flow. It also helps prevent shock, and protects the wire from corrosion and damage.

Insulation types are shown by letters stamped on the wire covering. Table 4-3 shows common wire insulations and approved locations. A complete listing is found in NEC Table 310-13.

Most wire is insulated with plastic or rubber. The insulation is treated during manufacturing to give it protective qualities. Special coverings such as asbestos and polyethylene are also available for special applications.

Certain types of insulation work better than others. They keep the wire cooler, reduce resistance and yield greater ampacity than a same size wire with less insulation.

Wires surrounded by air operate at lower temperatures than those enclosed in raceways or buried in the earth. Therefore, the NEC permits greater ampacities for wires in free air than for

Type	Insulation	Maximum Operating Temperature	Location
RH, RHH	Heat-resistant rubber	75°C./167°F.	Dry locations
RHW	Moisture and heat-resistant rubber	75°C./167°F.	Dry and wet locations
T	Thermoplastic; flame-retardant	60°C./140°F.	Dry locations
TW	Moisture-resistant thermoplastic; flame-retardant	60°C./140°F.	Dry and wet locations
THW	Flame-retardant; moisture and heat-resistant thermoplastic	75°C./167°F.	Dry and wet locations

Conductor Insulation
Table 4-3

enclosed ones, even though the wires are identical in size and insulation. See Table 4-4.

Locations
Depending on its insulation, wire is approved as suitable for wet or dry locations by the NEC.

A *dry location* is any place not *normally* subject to wetness or dampness.

A *wet location* is:

• An underground installation or one where the wire is embedded in a concrete slab.

• Any building subject to saturation by water or other liquids, either inside or outside.

• Any area unprotected against the weather.

The NEC may also require special wire types in *hazardous locations* (Art. 500, NEC) where flammable liquids, vapors or materials are present.

A *damp location* is also listed by the NEC. Areas such as canopies and roofed porches are damp locations. Inside rooms normally subject to moisture are also damp locations: a cold-storage warehouse and certain types of barns, for example.

Wires suitable *only* for dry locations cannot be used in damp or wet locations.

A wire suitable for a wet location may be used in all three locations, but their use is a *must* in a wet or damp location.

Using Insulated Wire
The NEC Table 310-13 lists over 30 types of insulation for wire sizes commonly used in electrical systems. Carefully examine each to determine the type that best fits the job at hand. You'll save time and money.

Size AWG	Three Conductors Or Less In* Raceways, Cables Or Earth	Single Conductors* In Free Air
14	15 amps	20 amps
12	20 amps	25 amps
10	30 amps	40 amps
8	40 amps	60 amps
6	55 amps	80 amps
4	70 amps	105 amps
2	95 amps	140 amps
0	125 amps	195 amps
00	145 amps	225 amps
000	165 amps	260 amps
0000	195 amps	300 amps

*Allowable Ampacities — Insulated copper conductors at 86° F.

Larger values are permitted, depending on type of insulation. Consult NEC Tables 310-18 & 19.

Conductor Ampacities
Table 4-4

For example, a wire commonly used in residential systems is Type T. It's a thermoplastic insulated wire which is flame-retardant and suitable for use where the maximum temperature won't exceed 140 degrees F. It is NEC approved for dry locations and may be used in all rooms in the house.

Type THW is also thermoplastic wire. Besides being flame-retardant, it's also heat and moisture resistant and so can be used in wet and dry locations. Since the insulation is heat resistant, it is approved for use to a maximum of 167 degrees F. This yields higher ampacity, which may be needed for certain circuits. Like Type T, Type THW can be used throughout the house. However, it may be more economical to use Type T for most of the work and Type THW only where its special properties are needed.

In certain areas of the wiring system, space may be limited. In this case, a wire such as THWN may be useful. THWN is identical to THW, except that its insulation is nylon. Nylon is a tougher, better insulator than thermoplastic. As a result, less thickness is required. Type THWN has all the qualities of THW but with a smaller diameter. It's also more expensive.

As you can see, there's a lot to consider when choosing wire. Every job is different. Select wire that meets code requirements and is economical without sacrificing practicality.

Cable

Cable is two or more wires grouped together with a common covering. Don't confuse the outer covering or sheathing on the cable with the insulation on the individual wires. Wires inside a cable must be insulated according to NEC regulations for individual wires. The outer covering or sheathing of the cable is designed to protect the wires inside. See Figure 4-5.

Cable is used in all types of wiring. It allows easier, faster installation in *open wiring* systems—when conduit is not required.

Cable is described by wire size and the number of wires in the cable. For example, a 12-2 cable contains two No. 12 wires, one black and one white. A 14-3 cable has three No. 14 wires, one black, one white and one red. The wires are color-coded to aid proper connection. A cable with a ground conductor will be labeled "w/g" or "w/gr." The ground is a bare (non-insulated) wire enclosed only by the cable covering. It's used as a grounding conductor in systems without conduit.

Besides the size and the number of wires, the insulation type is also stamped on the cable covering.

Other information, such as the maximum voltage of the cable, may also be included. A complete cable description may read: 12-2 NM 120 V. This description will be stamped every 24 inches along the cable covering.

Cable can also be a single group of stranded wires, No. 8 or heavier. Since stranded wire is more flexible, cable is generally preferred when larger diameter is needed. Current NEC regulations require that conductors No. 8 and larger be stranded when installed in raceways.

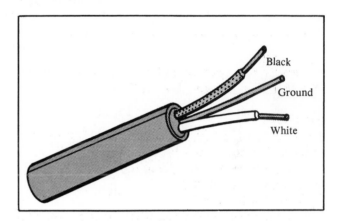

Typical Cable for Branch Circuit Wiring
Figure 4-5

Whether made up of a single-stranded conductor or a group of conductors, cable must be suitable for the location and the expected load. Ampacity is determined the same way as for single conductors. Suitability of location is determined by conductor insulation and the cable's outer covering.

For example, Types NM and NMC are non-metallic sheathed cable with two or more conductors. Both are moisture-resistant and flame-retardant. Type NM is approved for dry locations; NMC can be used in damp or corrosive areas. Both are used extensively in single- and multi-family dwellings that do not exceed three floors. They enclose 14-2 or 12-2 wire systems with or without ground and are used mostly for branch circuiting. The NEC permits Types NM and NMC for both exposed or concealed work. Often local codes will allow it only for concealed work. Check with your local inspector before installation.

Type UF is similar to NM and NMC in that it's used for branch circuiting in dwellings. However, UF is approved for direct burial underground and for wet locations.

Types SE and USE are service entrance cables. USE is suitable for underground use. Type SNM is a non-metallic sheathed cable used in hazardous locations. Type MI is a mineral-insulated metal-sheathed cable suitable for a wide range of applications and locations.

Type AC is *armored cable*. The insulated conductors are enclosed in a flexible metal tubing. Armored cable is sometimes called by its older name, BX cable.

Armored cable usually has two conductors, but three-wire cable is available. All armored cable must have an internal bonding strip of copper or aluminum. This strip runs the entire length of the cable and ensures that the cable will have a ground path.

AC cable is permitted by the NEC for branch and feeder circuits in both open and concealed work. Again, check your local codes. They may not allow it in open work. It can be embedded in brick or other masonry. If the location is damp or wet, Type ACL must be used. ACL has lead-covered conductors to protect it from moisture.

The use of armored cable in wiring systems has declined in recent years. Newer, flexible plastic cable is easier and cheaper to install.

Cords

Connecting a lamp, power tool or other appliance to a receptacle requires a *cord*. Cord consists of two or more insulated conductors with a common cover. Sometimes the conductors are stranded to give the cord greater flexibility.

Cord insulation is controlled by the NEC the same way as wire insulation. Qualities of the insulation determine whether the cord can be used in a dry or damp location, and the number and size of conductors the cord must have.

For example, Type S is a flexible cord that can be used on conductors from No. 18 to No. 2. It's suitable for damp locations and extra hard use. It's well insulated and found in both residential and commercial applications. Many washing machines, garage door openers and power tools such as grinders or lathes have Type S cord.

Type AFS is like Type S, but its insulation includes a layer of asbestos. This makes it suitable for electric heaters. Type C, on the other hand, has only a cotton insulation and is used only in dry locations with medium use. A table lamp cord would be Type C.

The NEC lists over 50 types of cord (Table 400-4). When installing new cord, be sure to use a cord with the right insulation and ampacity. And don't always replace an old cord with another of the same type. Many old cords are too small or can't handle the load placed. Determine how and

where the cord will be used; then choose the right type. Cords that are too small, of poor quality, and that are improperly used are one of the leading causes of electrical shock and fire.

Grounding Plug Adapter
Figure 4-6

Many appliances have cord with a green ground wire. This wire connects the housing of the appliance and the ground prong of the plug. When the appliance is plugged into the receptacle, the ground plug connects to the ground wire of the branch circuit. This keeps the appliance case grounded and prevents a shock if wiring inside the appliance fails. Most older electrical systems don't have receptacles with a ground slot. For these systems adaptors are available which permit the plug to be grounded to the outlet. See Figure 4-6. The "pigtail" is connected to a screw on the receptacle.

Color Coding of Wire

Wires are color coded, both for safety and to simplify installation. Never violate the color codes.

In residential and commercial systems, the neutral wire is white or grey. Hot wires are black, red, blue and yellow, depending on how many the circuit uses. Hot wires always follow this order in a correctly wired system.

In a few special cases, color codes are changed. In wiring a three-way switch, for example, the neutral wire is used as a hot wire. When this happens, the white neutral wire must be painted black where its ends connect to switches and outlets. This tells anyone working on the circuit that the neutral wire is really a hot wire.

Most ground conductors in open wiring systems are bare conductors enclosed in the cable cover.

These wires aren't needed in conduit because the conduit acts as the ground. Ground conductors in plugs are green or green with a yellow stripe; they should not be confused with the neutral wire. The differences are explained in the next section.

Many wires are striped in combinations of colors. These wires are used in low voltage installations, such as intercom and telephone systems. The distinctive color combinations aid in tracing and installation.

Ground Wires

It's easy to confuse the terms "grounded," "grounding" and "ground." Although very much alike, each word refers to a different part of the wiring system.

The white neutral wire is sometimes called the *grounded wire*. Normally the neutral wire doesn't carry any current. Its purpose is to act as a conductor to carry any unbalanced or unused current in the circuit back to the source. But it also has another purpose. Surges of voltage too large for the circuit such as lightning striking the system or a fault within the system are carried back to the protective devices by the neutral wire. The devices are tripped and excess voltage is drained off to the earth.

The *ground wire* connects the neutral wire and the earth. It's usually a No. 4 or larger bare conductor. One end is connected to the neutral wires in the service panel and the other end connected to a grounding electrode driven into the earth or attached to a metal water pipe. The neutral wires and their ground wire connection are generally called the *service ground*.

Besides the service ground, many wiring systems require an *equipment ground*. Equipment grounds are a fairly new NEC requirement. They were developed to control ground faults and short circuits. Service grounds are not sensitive enough to handle these shorts effectively.

Equipment grounds bond together conductors, enclosures and equipment. This is done with a *grounding conductor* or through metal conduit. The equipment ground protects anyone using the circuit from shock by providing a path back to the ground wire. When a fault occurs, the current follows this path rather than traveling through enclosures such as receptacles and appliance housings. Grounding installations are explained in detail in Chapter 7.

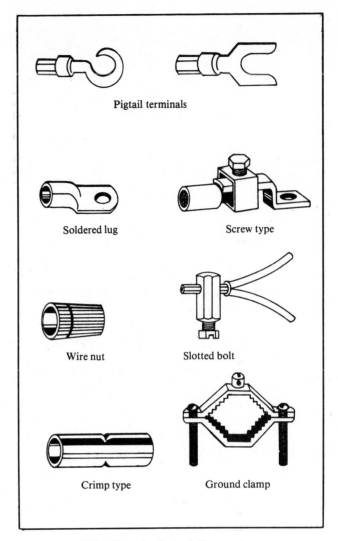

Wire Terminals and Connectors
Figure 4-7

Connections

Terminals

At the end of every wire there's a terminal. The simplest and perhaps most widely used terminal is made by stripping insulation from the end of the wire and wrapping the exposed wire around a screw. A terminal can also be made with a clip attached to the end of a wire. Most terminals are attached by pressure, although some must be soldered.

A crimp-type terminal has a shaft made of soft metal that's squeezed onto the end of a wire. Larger terminals use a screw-type lug, held in place by a screw or bolt. In some cases the lug is attached with a special indenting tool. Figure 4-7 shows several types of terminals.

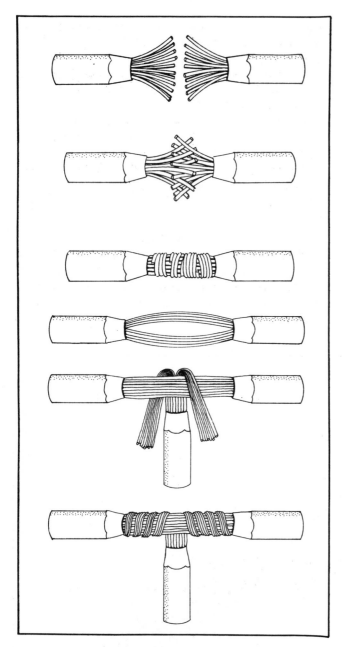

Splicing Stranded Wire Before Soldering
Figure 4-8

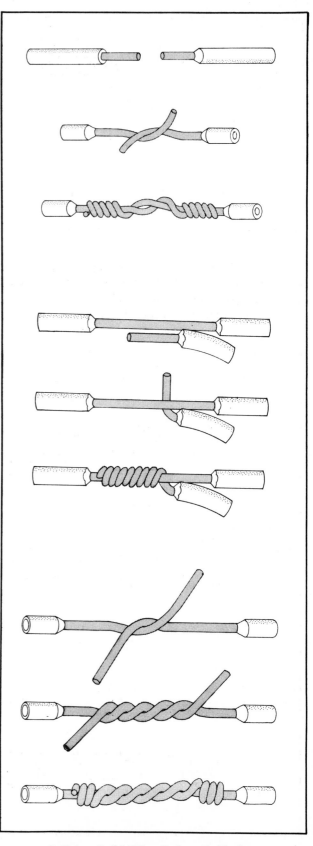

Splicing Solid Wire Before Soldering
Figure 4-9

Splices and Taps

Splices and taps are used to join conductors. They may be made with devices or by soldering.

Most splicing devices hold the wires by pressure. Wire nuts, spring connectors and screw connectors are used on smaller gauge wires. Large gauge wires use a split bolt or similar device.

Some splices and taps must be soldered. The NEC requires that the splices be mechanically and electrically secure before soldering. Common splices are shown in Figures 4-8 through 4-10.

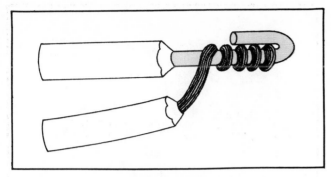

Splicing Solid to Stranded Wire
Figure 4-10

throughout the joint. Resin-type soldering paste (flux) must first be applied unless it's already included in the solder.

Use care when soldering. The heat from the gun should not damage wire insulation or other connections that are close by.

After the splice or tap is made, it must be insulated. Some pressure devices don't require insulation. But connections that do should be wrapped with electrical tape. The wrap should be equal or slightly heavier than the original insulation and be applied in neat spiral layers.

A method for splicing, soldering and taping wire is shown in Figure 4-11. Heat the wires with the soldering gun until solder melts when it touches wire. The solder should then flow smoothly

Conduit

Conduit is pipe which encloses and protects conductors. Not all parts of an electrical system use conduit. It's generally required to protect the wires in wet, corrosive or exposed locations. Commercial buildings usually are wired in conduit throughout

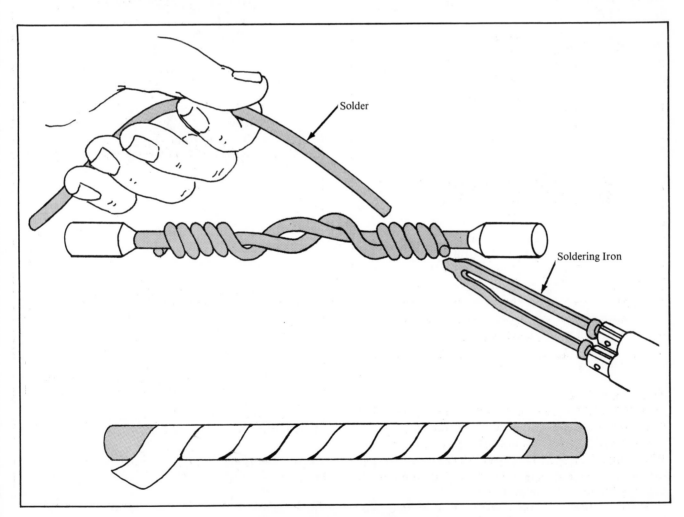

Solder

Soldering Iron

Splicing, Soldering and Taping Wire
Figure 4-11

for extra safety. Conduit is required in service areas of smaller systems where large voltages may present a danger if the wires are accidentally exposed.

Most of the early conduit installations used either rigid or thin-wall metal conduit. Although metal conduit is still used on many jobs, plastic-type PVC conduit has been approved for many applications. It's lightweight, corrosion resistant and easy to install. Many modern systems also use raceways. These are square or rectangular channels for wire that can be surface mounted on walls or placed flush in concrete floors. This makes adding or changing outlet locations easy—an important consideration in offices with many desks and appliances.

The NEC has grouped all enclosed channels for wire under the heading of "raceways." This includes conduit, raceways, busways and wireways. The following sections describe the types of conduit and other raceways commonly used in systems.

Rigid Metal Conduit

Rigid conduit is a thick-walled pipe that comes in ten-foot lengths with either coated black enamel or galvanized finish. It's ideal for locations where wire must be protected from wetness, corrosion or physical damage.

Galvanized conduit is used outdoors and in areas subject to moisture. It can also be buried in earth or concrete.

Conduit can't be buried in cinder fill where moisture is present unless additional protection is used. Cinder fill is highly corrosive to steel.

Rigid conduit is threaded at the factory. It's joined with connectors and couplings in the same manner as water pipe. (Figure 4-12 shows typical conduit connections and couplings.) The conduit can be cut, bent, and threaded in the field. Approved watertight connectors and couplings must be used in concrete, masonry or outdoor work.

Small diameter conduit is bent by hand with a tool called a *hickey*. (See Chapter 3.) Large diameter conduit must be bent with a power bending tool.

Rigid Non-metallic Conduit

Non-metallic conduit can be made of polyvinyl chloride (PVC), fiberglass epoxy or similar material. It is lightweight and easily assembled using couplings and bonding cement. It can be used in most locations, is very good at resisting corrosion, and can be buried in earth, cinders or concrete. In wet or damp locations, the connections must be watertight.

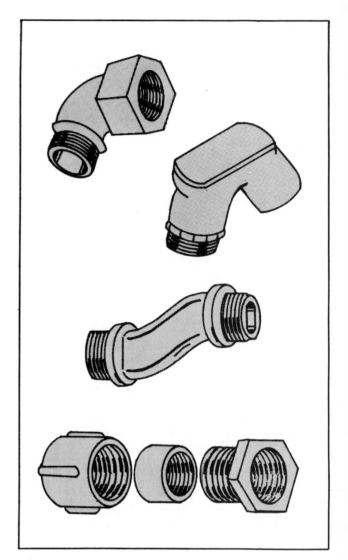

Rigid Conduit Connections and Couplings
Figure 4-12

Non-metallic conduit can't be used in hazardous locations where fumes or combustible dust might cause a fire or explosion. It can't be used to support fixtures or equipment. Although strong, it doesn't resist impact as well as rigid metallic conduit.

Non-metallic conduit is made of plastic, so it can't be used where temperatures will be high. Another disadvantage is that non-metallic conduit can't conduct current. Unlike metal conduit, it must have a ground conductor wire where grounding is required.

Non-metallic conduit is bent with a special tool that heats it to the right temperature. It's easy to bend it into shape when hot. The conduit cools and hardens quickly and permanently. Factory-bent sections are also available.

Electrical Metallic Tubing

Electrical metallic tubing is also called EMT or thin-wall. It can be used indoors and out, in many locations. In wet locations or when buried in earth or concrete, it must be protected the same as rigid metallic conduit. It's not as strong as rigid conduit and should not be used where it would be subject to severe physical stress.

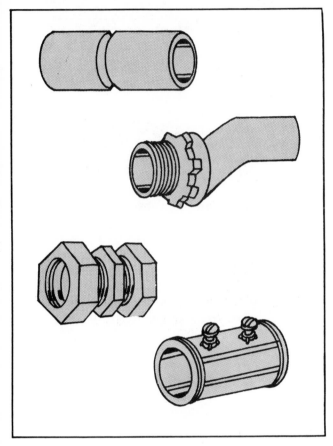

EMT Connections and Couplings
Figure 4-13

EMT is quicker to work with than rigid metallic conduit. It's lighter and easier to cut and bend. Connections are made with a special tool called an indenter. Compression couplings are also used (Figure 4-13). See the section on conduit tools in Chapter 3.

Intermediate Metal Conduit

Intermediate metal conduit is sturdier than EMT but not as heavy as rigid metal conduit. The NEC rules for use and installation are similar to those for rigid metal conduit and are outlined under Article 345. Intermediate metal conduit costs less than rigid and you may be able to save some money on some jobs. Be sure to check local codes and job specifications, since many times they'll reject the use of intermediate conduit.

Installation of intermediate conduit is done in the same way as the rigid conduit installation steps described in Chapter 7.

Flexible Conduit

Both EMT and conduit are available in flexible form. Flexible conduit and tubing are closely related to armored cable. In practice, it's the sheath of armored cable without the conductors inside. After it's installed, the wires are pulled through the same way as in other conduit. Flexible conduit is sometimes used to upgrade armored cable systems. It's often used to make equipment connections where vibration makes rigid conduit unusable.

Regular flexible conduit can be used only in dry locations. Special watertight flexible conduit (usually called greenfield) can be buried in earth or concrete and used in wet or hazardous locations. No flexible conduit can be used where there's a chance of damage from impact. Thin-wall flex is not permitted in lengths over six feet.

Restrictions

The following general rules apply to all conduit:
• Depending on type and size, all conduit must be supported at minimum distances by straps or hangers. See Table 4-14.
• No size smaller than 1/2 inch is permitted except in special cases.
• Whether flexible or rigid, not more than four bends totaling 360 degrees are permitted in any one run.
• Depending on the diameter and type of conduit, the distance (radii) of the conduit bends is also limited. See Table 4-15.
• Wire splices and taps may be made only in junction boxes, outlet boxes or at a conduit body. A conduit body is a separately installed part of the conduit system with a removable cover. It can be used at a junction of two or more sections of conduit or at a terminal. The conduit body must be at least twice the cross-sectional area of the largest conduit connected to it. Conductors, splices and taps can't fill the conduit body to more than 75% of capacity.
• When conduit is cut, the pipe end must be reamed or filed to protect wires from damage when they're pulled through.
• To protect the wires where conduit enters a box or other fitting, a bushing or adapter must be used

Rigid Metal		Rigid Non-metallic	
Conduit Size (Inches)	Maximum Distance Between Supports (Ft.)	Conduit Size (Inches)	Maximum Distance Between Supports (Ft.)
½ - ¾	10	½ - 1	3
1	12	1½ - 2	5
1¼ - 1½	14	2½ - 3	6
2 - 2½	16	3½ - 5	7
3 and larger	20	6	8
EMT/Thinwall		Flexible Metal Conduit	
All sizes	10	All sizes	4½*

*Except where fished.

All solid conduit must be fastened within 3 feet of each outlet box, junction box, cabinet or fitting.

Flexible conduit must be fastened within 1 foot of each junction, outlet or fitting.

Maximum Distances for Conduit Supports
Table 4-14

unless the box is designed to give protection by itself.

Conduit size is is based on the inside diameter. Sizes are given in inches: 1/2", 3/4", 1", 1¼", 1½", 2", 2½", 3", 3½", 4", 4½", 5", and 6" are typical conduit sizes.

Besides the general rules mentioned above, the NEC limits the number of conductors in conduit. This prevents overheating that could result if too many wires or if too large a wire were enclosed. It also prevents wires from damage which can occur when too many are pulled through the conduit. Table 4-16 shows the conduit size requirements for the most common conductor sizes.

Refer to Chapter 9 of the NEC, Tables 4 through 8 if you plan to run conductors of different sizes in the same conduit. These tables give conduit sizes in cross-sectional area and list the permissible percentage of cross-sectional area that the conductors

may take-up. Tables listing conductor cross-sectional area can then be used to make the necessary calculations.

For example, let's say a job requires three No. 14 conductors and two No. 12 conductors run in 3/4-inch conduit for maximum economy. Is this permissible?

From NEC Table 4, in Chapter 9, we find that 3/4-inch conduit has a total area of 0.53 square inches. For more than 2 conductors, 40% or 0.21 square inches can be used. Table 5 shows Type RHH wire as having an area of 0.0327 square inches. No. 12 of the same type is 0.0384 square inches.

No. 14 3 x .0327 = .0981
No. 12 2 x .0384 = .0768
.1749 total area for all conductors

Conduit Size (Inches)	For Conductors w/o Lead Sheath (Inches)	For Conductors w/Lead Sheath (Inches)
½	4	6
¾	5	8
1	6	11
1¼	8	14
1½	10	16
2	12	21
2½	15	25

Allowable Radius of Conduit Bends
Table 4-15

Conductor Size AWG	Number of Conductors			
	1	2	3	4
	Conduit Size Required (Inches)			
14	½	½	½	½
12	½	½	½	½
10	½	½	½	½
8	½	½	½	¾
6	½	¾	1	1
4	½	1	1	1½
3	½	1	1¼	1¼
2	½	1	1¼	1¼
1	¾	1¼	1¼	1½
0	¾	1¼	1½	2
00	¾	1½	1½	2
000	¾	1½	2	2
0000	1	2	2	2½

For additional sizes and/or number of conductors, consult the NEC Chapter 9, Tables 1 through 3C.

Conduit Size Requirements
Table 4-16

Since 0.1743 is less than 0.21, the five conductors are O.K.

Raceways, Wireways and Busways

As mentioned in the section on conduit, the NEC defines any enclosed channel designed to hold wires as a raceway. Technically, conduits are raceways. As commonly used, however, raceways are square, rectangular or circular metal enclosures other than conduit.

A raceway often found in electrical remodeling is called *wire molding* or *wire mold*. This is a small rectangular metal raceway designed to supply the wires to a receptacle or switch. It must be used only in dry, exposed locations. Wire molding is available in a variety of colors to match or complement room colors. It can also be painted.

Most raceways, however, are designed to hold the large number of wires needed in commercial systems. The NEC permits up to 30 conductors in any one channel. There may be more than one channel in a raceway. This permits the separation of different circuits, when required by the NEC. Raceways may be surface mounted or installed flush with the floor. Some raceways become part of the building's floor slab or metal floor.

Because raceways carry wire like conduit, they have to follow the same rules on installation, location, size and number of conductors.

Wireways are like raceways, except that one side is hinged or removable so the wire can be laid in rather than pulled in place. As a general rule, wireways are used to hold wire over shorter distances than raceways.

Wireways can be used only for exposed work. No more than 30 conductors are permitted. The sum of all conductor cross-sectional area can't be more than 20% of the interior area of the wireway.

A *busway* is a metal enclosure containing factory mounted conductors. These conductors may or may not be insulated and are usually copper rods or bars. Busways are often found in service and feeder equipment.

Only approved raceways, wireways or busways can be used in a damp or wet location. Other NEC requirements are listed under the appropriate sections in the Code Handbook.

Messenger-Supported Wiring

Messenger-supported wiring uses a suspended high-strength wire to support the insulated conductors. The support wire is sometimes separately strung and uses rings or saddles to carry the conductors. The support wire also may be a bare wire twisted with the insulated conductors in a factory-made assembly. Support clips are used to attach it to the pole or building.

Messenger wire systems can be used in dry or hazardous locations if the conductors are the proper size and type. Messenger systems are common in factories, outdoor lighting systems and service drops.

Outlet, Pull and Junction Boxes
Outlet Boxes

Outlet boxes tie together the different sections of the electrical system. They hold conduit or cable in position. In conduit systems, they are part of the continuous ground path. Most important, they provide space for mounting and connecting devices such as switches, fixtures and receptacles.

Boxes are available in steel, PVC, aluminum or cast iron. Common shapes are square, round, rectangular, or octagonal.

A wide range of boxes is available. Some of the more common types are shown in Figures 14-17 through 14-19. Many boxes have one side that can be removed so the box can be joined or "ganged" to another when needed. Some boxes have brackets for quick mounting. Others have special expansion-type mountings that install in an existing wall or ceiling with a minimum of sawing or chipping. Most boxes have knockout plugs for conduit

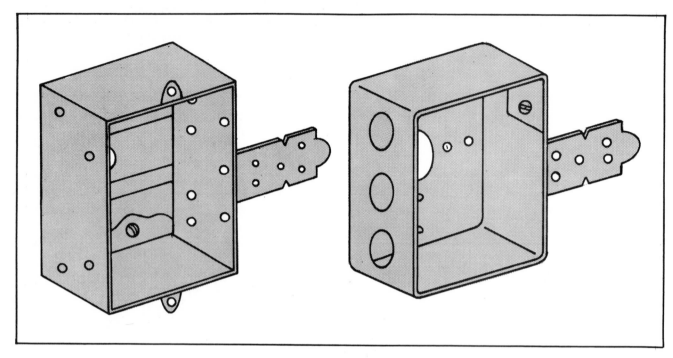

Typical Outlet Boxes
Figure 4-17

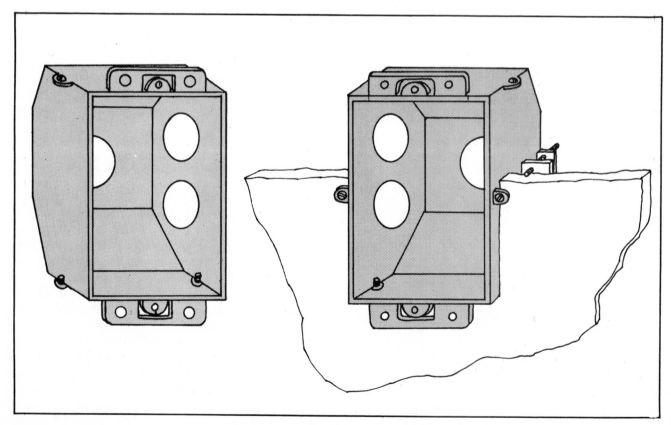

Outlet Boxes for Old and New Work
Figure 4-18

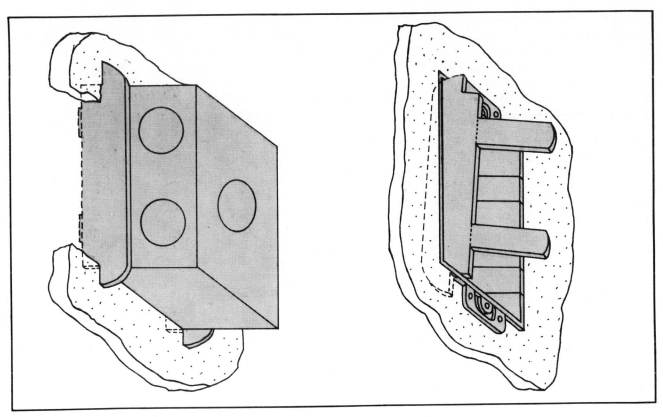

Outlet Boxes for Old and New Work
Figure 4-18 (continued)

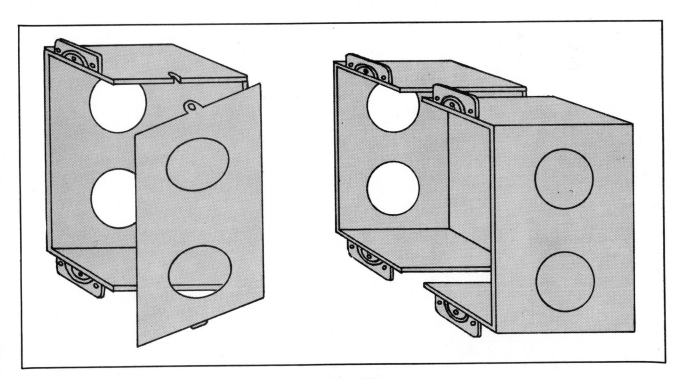

Ganging Metal Boxes
Figure 4-19

and cable, and many have cable clamps inside to secure the cable.

Your choice of an outlet box depends on the location and how many conductors it must hold. The NEC requires a certain amount of space for each conductor. See Table 4-20. The following rules also apply:

• Any conductor in the box which doesn't enter or leave the box is not counted.

• Any conductor running *through* the box is counted as one conductor.

• Any conductor ending in the box is counted as one conductor.

• When one or more *grounding* conductors enter the box, deduct one live conductor.

• Items such as fixture studs or cable clamps require a deduction of one conductor for *each type* of device in the box.

For example, a 4" x 1½" round box can hold seven No. 14 wires. If the box must also hold two cable clamps and one fixture stud, two conductors would have to be deducted—one for each device. The maximum number of conductors permitted would be 5.

Boxes come in standard depths of 1¼ inches, 1½ inches, 2⅛ inches and 3½ inches. For special jobs, boxes as thin as 1/2 inch can be used. No outlet box can be less than 1/2 inch. The greater the depth, the more space for conductors. It also makes splicing and pulling easier. So use the largest size box that will fit in the wall, if location and job cost permit it.

Use steel or cast iron boxes for rigid or thin-wall conduit or armored cable. These boxes are galvanized or coated with enamel. Galvanized boxes are required in wet locations and must be mounted with at least 1/4 inch space between the box and the mounting surface. They must also be watertight. Boxes designed for armored cable or flexible conduit have screw clamps mounted inside next to the knockouts. This eliminates the need for cable connectors, reducing installation time and cost.

Steel boxes are also used with non-metallic cable in open wiring. Box openings and clamps have a smooth surface to prevent damage to the cable.

Boxes made of PVC are used with PVC conduit. PVC and bakelite boxes are also acceptable for cable systems (Figure 4-21), but aren't permitted in metal conduit systems because they can't act as a ground. They're installed the same way as steel boxes.

Mount outlet boxes so they are secure and rigid. If you can't mount them to a stud or wall, use bar hangers or sturdy wooden cleats. Boxes with an interior volume of less than 100 cubic inches can be supported by two or more lengths of conduit. The conduit must be tight to the box and properly supported within 18 inches of the box.

Type	Size (Inches)	Maximum # of Conductors				
		#14	#12	#10	# 8	# 6
Round/ Octagonal	4 x 1¼	6	5	5	4	0
	4 x 1½	7	6	6	5	0
	4 x 2⅛	10	9	8	7	0
Square	4 x 1¼	9	8	7	6	0
	4 x 1½	10	9	8	7	0
	4 x 2⅛	15	13	12	10	6
	4¹¹⁄₁₆ x 1½	14	13	11	9	0
	4¹¹⁄₁₆ x 2⅛	21	18	16	14	6
Device	3 x 2 x 2	5	4	4	3	0
	3 x 2 x 2½	6	5	5	4	0
	3 x 2 x 3½	9	8	7	6	0
	4 x 2⅛ x 1⅞	6	5	5	4	0
	4 x 2⅛ x 2⅛	7	6	5	4	0

For additional sizes, consult NEC Table 370-6.

Maximum Conductors in Outlet Boxes
Table 4-20

Stud-mounted PVC Box in Cable System
Figure 4-21

After the wiring job is done, all boxes must be covered with a face plate or fixture. Any unused openings must be plugged.

Pull and Junction Boxes

In some installations, boxes are used primarily for splicing and pulling wires. Where the wires are No. 4 or larger, the boxes must be large enough so wire can be pulled through without damage.

In straight pulls, the length of the box can't be less than 8 times the diameter of the largest raceway. If the pull is U-shaped or at an angle, the distance between each raceway entering the box and the opposite side of the box can't be less than 6 times the diameter of the largest raceway.

Boxes smaller than those mentioned above can be used for No. 6 and smaller conductors. The size of these boxes must follow the rules given for outlet boxes in the previous section. The NEC requires that they be marked with the maximum number and size of conductors they can carry.

All pull and junction boxes must have a cover. After installation, the box must be accessible for future work.

Switches, Receptacles and Fixtures
Switches

The NEC classifies push button and toggle switches as *snap switches*. See Figure 4-22. A simple toggle switch clicks on and off by moving a lever up or down or side to side. The clicking sound is made by a spring mechanism inside the switch. Silent toggle switches have a mercury contact. A push button switch operates on the same mechanical principle as a toggle switch.

Many types of snap switches are available. Some just turn the power on and off. Others are designed for certain jobs. See Figure 4-23.

Some switches combine several functions. A rotary dimming fixture switch combines a rheostat with push-button action. This allows raising or lowering the light level of incandescent fixtures as well as on and off action. Some switches have several positions which can control the speed of a motor. Some switches have tiny lights in the handles which glow when the switch is off or on. Weatherproof switches are available for use in wet locations. See Figure 4-24.

Single circuits with a heavy load, such as a large motor, use heavy-duty switches as shown in Figure 4-25. These switches usually have an overload protection device which acts like a circuit breaker. A reset button closes the breaker when the motor has

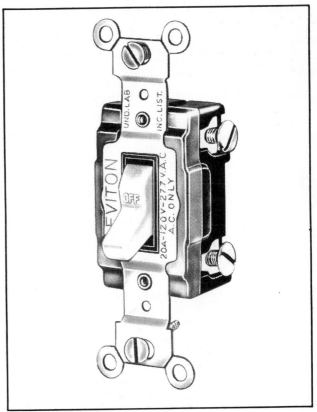

Courtesy: Leviton Manufacturing Co.

Household Snap Switch
Figure 4-22

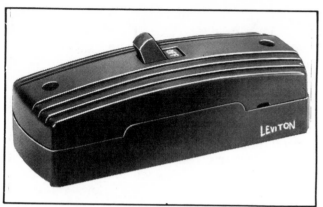

Courtesy: Leviton Manufacturing Co.

Snap Switch used with Surface Wiring
Figure 4-23

cooled so the circuit can return to operation. For some simple circuits, the circuit breaker itself is the switch. More information on motor switching can be found in Chapter 6.

Service equipment also uses heavy-duty switches. Depending on the design and installation, a snap switch or a knife switch may be used.

A *knife switch* has blades which slide between clips to make contact. The blades are operated by a lever. Knife switches have cartridge or plug fuses to protect against overload. Switches rated at over

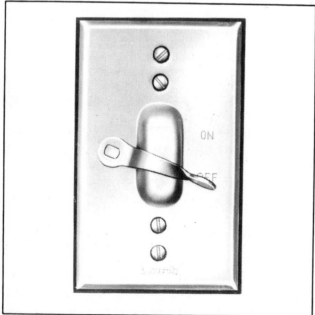

Courtesy: Leviton Manufacturing Co.

Weatherproof Switch
Figure 4-24

1200 amps at 250 volts or less, or at over 600 amps at 251 to 600 volts, can't be used to open the circuit under load because of the possibility of shock. Instead, use a circuit breaker or a switch designed for the load. Then use a knife switch to isolate the circuit or equipment from the power source.

Knife and snap switches used for situations other than the high ampacities listed above are classified as general-use switches. General-use switches are

Courtesy: Square D Company

Heavy Duty Motor Switch
Figure 4-25

further broken down by the NEC into AC and AC/DC. An AC general-use snap switch can be used under the following conditions:

• For resistive and inductive loads, the loads can't exceed the ampere rating of the switch at the applied voltage.

• Tungsten lamp loads can't exceed the ampere rating at 120 volts.

• Motor loads can't exceed 80% of the ampere rating of the switch at its rated voltage.

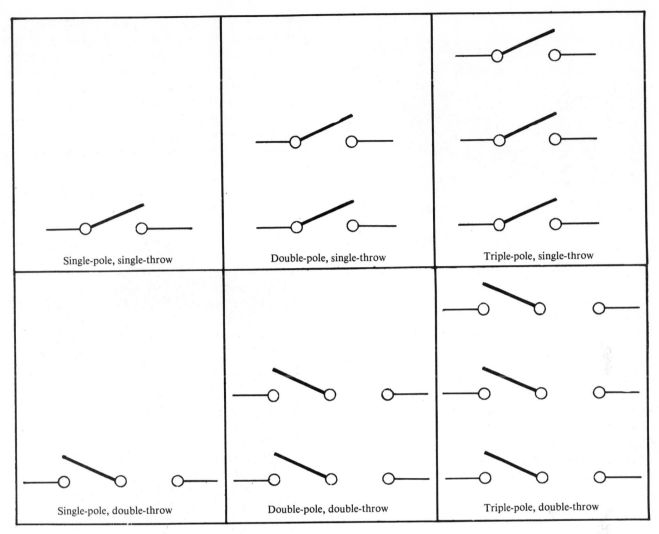

Circuit Diagrams of Pole Switches
Figure 4-26

An AC/DC general-use snap switch can be used under the following conditions:

• For resistive loads, the load can't exceed the switch ampere rating at the applied voltage.

• Inductive loads can't exceed 50% of the switch ampere rating at the applied voltage.

• Tungsten lamp loads can't exceed the switch ampere rating at the applied voltage. A "T"-rated switch is required.

Switches are also classified by their use in the circuit, as diagrammed in Figure 4-26.

A single-pole, single-throw (SPST) switch operates in one side of the line. (Figure 4-27). It interrupts or connects one conductor.

A single-pole, double-throw switch (SPDT) permits a conductor to be connected to either of two other conductors.

A double-pole, single-throw (DPST) switch operates in both sides of the line. It interrupts or connects two conductors.

A double-pole, double-throw (DPDT) switch allows a pair of conductors to be connected to either of two other pairs.

Triple-pole, single- and double-throw switches are also used. They're found mostly in three-phase wiring.

The switch used depends on the installation. For example, three- and four-way switching will use SPDT and DPDT switches.

Two or more SPST switches installed in one box under one faceplate is called a *gang switch*. Each switch controls a separate fixture or appliance.

Switch location must also be considered. The National Electrical Manufacturers' Association (NEMA) has developed a rating system for switch construction. A NEMA Type 1 switch is suitable

SPST Switch and Hot Wire
Figure 4-27

for indoor use. The 3R Type is rainproof. Type 4 is waterproof and dustproof. Types 7 and 9 are approved for hazardous locations. Other types and applications are found in the manufacturers' catalogs.

In almost all circuits, switches are wired only to interrupt current flow in the hot wire or wires—the black and red. The white wire—the grounded conductor—and the bare ground wire are never wired to the switch, except when the switch is used to isolate a motor or when the white wire is used as a hot wire for 3-way switching. Chapters 6 and 7 have information on wiring switches into different circuits.

Fixtures

A *fixture* is a lighting device connected directly to the circuit. A *lampholder* is a simple lighting fixture with a socket for a lamp (bulb) and maybe a pull chain. See Figure 4-28. When working with lighting systems, the proper term for a lighting fixture is a *luminaire*. What is commonly called a bulb is technically a *lamp*. Chapters 5 and 6 have a detailed explanation of fixtures and lighting systems.

Receptacles and Plugs

Portable appliances and devices are connected to a circuit through an outlet called a *receptacle*. The most common are single and duplex (double), as

shown in Figure 4-29, with three-wire terminals. Two of the terminals connect the hot and neutral wires. The third is the grounding terminal. Receptacles are designed to receive plug prongs either by straight pushing or by twist-and-turn pushing.

The duplex receptacle is most common, but many other shapes are in use. Each is used on a circuit of a certain ampacity. The shape and design of the receptacle keeps anyone from plugging in an appliance or device rated at a lower amperage. This prevents damage to the device and shock to the user. Plug and receptacle configurations are shown in Figure 4-30.

The terminals in plugs and receptacles are marked to ensure that the proper conductor is wired to the correct terminal. The letter G identifies the terminal for attaching the *grounding* wire of a device or appliance. The terminal marked W is for the *grounded* circuit wire. This is the white or gray wire. The live or "hot" conductor is connected to the unmarked terminal. Where there are two or three terminals for hot wires, they are marked X, Y and Z.

Since 1980, the NEC has required that many portable appliances have a polarized plug. A *polarized plug* has two prongs, one wider than the other. With a polarized plug, the switch on the appliance will always turn off the hot line in the circuit. With a non-polarized plug, the switch may become part of the neutral line. This will still turn the appliance

Courtesy: Leviton Manufacturing Co.

Porcelain Lampholder
Figure 4-28

Courtesy: Leviton Manufacturing Co.

Single and Double Receptacles
Figure 4-29

---

on and off, but the appliance may remain hot, even with the switch in the off position.

A plug with more than two prongs is always polarized.

Fuses and Circuit Breakers

Fuses
Fuses are one of two overcurrent protection devices in common use. They open the circuit automatically when the current rises above a safe limit. Fuses rely on heat. When the heat in the conductor rises, the element in the fuse melts, interrupting the circuit.

There are two types of fuses: plug fuses and cartridge fuses (Figure 4-31). A *plug fuse* screws into a base and is used in circuits with 125 volts or less. Plug fuses have a transparent top which lets you see if they're blown. They come in 10-, 15-, 20- and 25-amp ratings.

Type S fuses prevent deliberate or accidental replacement of low-current fuses with high-current fuses. They have a special thread size which makes it impossible to install them in a conventional fuse base without using an adaptor in the base that is hard to remove.

Special Type S fuses can be bought with either a time-delay or slow-blow element. A time-delay fuse has an element which will not blow under a short, heavy ampere drain. Demands of this type occur in motor circuits whenever the motor is started. Under short circuit conditions, however, the fuse will blow.

Circuits that draw 30 or more amps use *cartridge fuses*. Cartridge fuses slide into clips rather than screw-type bases. Some use blade or rod contacts. These fuses are usually located in a service box with a disconnect switch.

Cartridge fuses have different lengths and thicknesses for different amp ratings. This prevents substituting the wrong size. They're also available in time-lag design.

Some cartridge fuses can't be visually inspected. A test lamp is needed to see if they're blown. Others, however, have a renewable element. You can take the fuse apart and if the element is melted, insert a new one.

Circuit Breakers
Circuit breakers are like an automatic snap switch. When current in a circuit exceeds safe limits, the breaker trips, cutting off the current flow. They're also available with time-delay features. Figure 4-32 shows single-, double-, and triple-pole circuit breakers.

Circuit breakers have a big advantage over fuses since they can be reset. Blown fuses must be replaced. Breakers can also act as switches in the circuit for some installations. And breakers are easier to install. Most breakers snap in and out of the service panel. Because of these advantages, breakers have replaced fuses in most modern wiring system.

Breakers are thermal, magnetic or a combination of the two. A *thermal circuit breaker* has a bimetallic element that reacts to changes in temperature. The extra heat in a shorted element trips the breaker. A *magnetic circuit breaker* reacts to changes in the current flow. An increased current flow creates enough magnetic force to "pull up" an armature, opening the circuit.

A thermal-magnetic breaker combines features of both. Of the three types, the magnetic breaker may have an advantage in some locations, since it is not affected by temperature.

Breakers are sized by the maximum amperage they can control and by the number of poles or wire connections each breaker has. For example, a 20-amp single-pole breaker would be suitable for a 2-wire, 20-amp circuit. The hot wire would be connected to the breaker pole and the neutral wire to the neutral bar in the service panel. A 3-wire circuit

NEMA CONFIGURATIONS FOR GENERAL-PURPOSE NONLOCKING PLUGS AND RECEPTACLES

		15 AMPERE		20 AMPERE		30 AMPERE		50 AMPERE		60 AMPERE	
		RECEPTACLE	PLUG	RECEPTACLE	PLUG	RECEPTACLE	PLUG	RECEPTACLE	PLUG	RECEPTACLE	PLUG
2-POLE 2-WIRE	1 · 125 V	1-15R	1-15P								
	2 · 250 V		2-15P	2-20R	2-20P	2-30R	2-30P				
	3 · 277 V				(RESERVED FOR FUTURE CONFIGURATIONS)						
	4 · 600 V				(RESERVED FOR FUTURE CONFIGURATIONS)						
2-POLE 3-WIRE GROUNDING	5 · 125 V	5-15R	5-15P	5-20R	5-20P	5-30R	5-30P	5-50R	5-50P		
	6 · 250 V	6-15R	6-15P	6-20R	6-20P	6-30R	6-30P	6-50R	6-50P		
	7 · 277 V AC	7-15R	7-15P	7-20R	7-20P	7-30R	7-30P	7-50R	7-50P		
	24 · 347 V AC	24-15R	24-15P	24-20R	24-20P	24-30R	24-30P	24-50R	24-50P		
	8 · 480 V AC				(RESERVED FOR FUTURE CONFIGURATIONS)						
	9 · 600 V AC				(RESERVED FOR FUTURE CONFIGURATIONS)						
3-POLE 3-WIRE	10 · 125/250 V			10-20R	10-20P	10-30R	10-30P	10-50R	10-50P		
	11 · 3 ø 250 V	11-15R	11-15P	11-20R	11-20P	11-30R	11-30P	11-50R	11-50P		
	12 · 3 ø 480 V				(RESERVED FOR FUTURE CONFIGURATIONS)						
	13 · 3 ø 600 V				(RESERVED FOR FUTURE CONFIGURATIONS)						
3-POLE 4-WIRE GROUNDING	14 · 125/250 V	14-15R	14-15P	14-20R	14-20P	14-30R	14-30P	14-50R	14-50P	14-60R	14-60P
	15 · 3 ø 250 V	15-15R	15-15P	15-20R	15-20P	15-30R	15-30P	15-50R	15-50P	15-60R	15-60P
	16 · 3 ø 480 V				(RESERVED FOR FUTURE CONFIGURATIONS)						
	17 · 3 ø 600 V				(RESERVED FOR FUTURE CONFIGURATIONS)						
4-POLE 4-WIRE	18 · 3 ø 208Y/120 V	18-15R	18-15P	18-20R	18-20P	18-30R	18-30P	18-50R	18-50P	18-60R	18-60P
	19 · 3 ø 480Y/277 V				(RESERVED FOR FUTURE CONFIGURATIONS)						
	20 · 3 ø 600Y/347 V				(RESERVED FOR FUTURE CONFIGURATIONS)						
4-POLE 5-WIRE GROUNDING	21 · 3 ø 208Y/120 V				(RESERVED FOR FUTURE CONFIGURATIONS)						
	22 · 3 ø 480Y/277 V				(RESERVED FOR FUTURE CONFIGURATIONS)						
	23 · 3 ø 600Y/347 V				(RESERVED FOR FUTURE CONFIGURATIONS)						

Plug and Receptacle Configurations
Figure 4-30

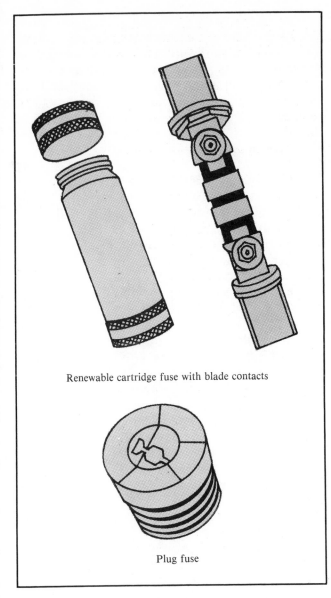

Renewable cartridge fuse with blade contacts

Plug fuse

The Two Types of Fuses
Figure 4-31

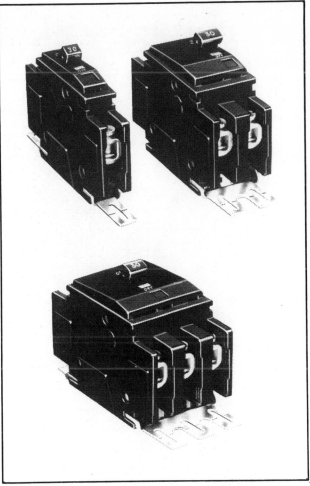

Courtesy: Square D Company

Single-, Double-, and Triple-Pole Circuit Breakers
Figure 4-32

would need a double-pole breaker. A 4-wire circuit would use a triple-pole type. Many amp and pole combinations are available to suit circuit requirements.

All ungrounded conductors in a circuit must be protected by an overcurrent device. The fuse or breaker is installed at the point where the conductor gets its power supply. Chapter 6 explains what the NEC requires when sizing overcurrent devices. Trouble-shooting fuses is explained in Chapter 8.

Service Equipment

Service equipment in the electrical system includes panelboards and load centers, switches, meters,

overcurrent protection devices and the necessary enclosures and hardware for these items. Most of this equipment is located where the main power is distributed into the branch or feeder circuits.

Service equipment has several purposes. It connects conductors with circuit breakers or fuses as well as to the main power. It also provides a safe, neat and practical means of installing and routing power into the various circuits. The breakers and switches protect the system and permit separate disconnection of circuits.

Metal enclosures protect connections against physical damage, moisture and corrosion. They also safeguard against shock and anchor the conduit. Meters record the amount of power used by the whole system or by individual circuits.

Most single-family homes need only one panel (Figure 4-33) to handle all of these functions except metering. The main disconnect and all breakers

200 Amp Service Panel/Load Center
Figure 4-33

will be in one enclosure. In larger installations, functions may be divided into separate panels and switches. This avoids having a single panel that's overcrowded and hard to work with. In multi-family dwellings, separate panels and meters are needed for separate meters and switching for each unit. In commercial buildings, separately controlled and metered branch or feeder circuits may be more efficient and convenient. Figure 4-34 shows a meter panel for a multiple unit dwelling where each apartment occupant is billed separately for electricity.

Service panels, also called load centers, have a cover and a faceplate. The cover protects against accidental contact and the weather. The faceplate, located behind the cover, contains the breaker switches and a chart listing the circuit numbers and locations. Removing the faceplate will expose the body of the panel. The body has the mechanical connections which wire the various circuits.

Terminals connecting the hot wires to circuit breakers or fuses are in vertical rows. The breakers

or fuses connect to the power source. The panel will have a *neutral bar* for grounding conductors. This bar connects the grounded conductor and the grounding conductor with the ground wire.

Service equipment should be placed in a building as close to the area of heaviest use as practical. This reduces voltage drop. If the equipment has more than one panel, the following sequence is common:

1. Service drop/entrance wiring
2. Meter
3. Main switch/overcurrent protection
4. Branch/feeder circuits

The exact position will depend on utility requirements, space available for mounting and the equipment used. It's best to follow a left-to-right arrangement for the cabinets.

The utility company uses a watt-hour meter to measure the power used by the customer. The meter is almost always furnished by the utility; its socket and necessary wiring is done by the customer. Depending on utility rules, it may be mounted inside the building with the other service equipment, or be placed outside the building for convenient reading. A minimum mounting height is required. Meters and their sockets must be of the rated ampacity for the service they meter.

Other service equipment is also located outside the building. The service drop—the wires from the utility's line—is anchored to the side of the building by a *house rack* or *service insulators*. The heavy conduit that runs up the side of the building between the meter and the service drop is called the *mast*. The watertight cap on its top end is called the *entrance cap, rain cap* or *mast head*. Service installations are explained in Chapter 7.

Check local building codes to see if they affect the location of service equipment inside the building. All equipment must be mounted plumb and level. Damp or wet locations require watertight equipment *and* a 1/4-inch space between the panels and the mounting surface. In some commercial installations, the NEC requires that service equipment be located in a room or vault. With few exceptions, no other piping, ducts or equipment are permitted in the same room. The rooms must be at least 6 feet 3 inches high.

For these installations, and all others as well, a minimum of three feet must be kept clear in front of the service equipment. (Service equipment in dwelling units using 200 amps or less is not included.) In areas where the work space has live conductors on both sides, 4 feet is required.

Work areas must have good lighting. Equipment that needs air flow to remove heat must be installed so the air flow will not be blocked.

Courtesy: Square D Company

Meter Panel for a Multiple Unit Dwelling
Figure 4-34

The NEC also regulates the ampacity of service equipment and the types of breakers and mechanical disconnects. These are explained in Chapters 6 and 8.

Manufacturers of equipment such as air conditioners, commercial hot water heaters and similar equipment sometimes sell pre-wired equipment panels designed specifically for that piece of equipment. These panels require little work other than mounting and connecting the main wiring to a proper-size breaker. Switches and circuit breakers are included in the panel. These panels save time and will usually be worth the extra cost.

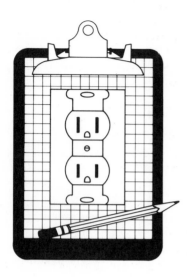

Chapter 5:

Planning Electrical Systems

When you plan or lay out an electrical system, there are three key areas: service, lighting and receptacles. Service includes the service entrance, metering, and switchgear. Lighting is the circuits and fixtures needed for illumination. Receptacles provide power at points where appliances and equipment will be used.

Of course, these three areas overlap to a certain degree in most buildings. For example, a 20-amp branch circuit in a house will generally supply both lighting and receptacles. Still, we'll avoid confusion if we consider each area separately.

Why Design is Important to You
On many jobs, especially larger commercial and industrial jobs, the electrical system plan will be drawn up by a design firm at the expense of the owner. In this case, your task is to study the plan, submit a bid and then follow the plan, job specs and the code if you get the contract.

For many jobs, however, there won't be a plan. It will be up to you to draw it up. Your ability to do this will save you the money and expense of hiring a professional designer. However, be aware that for some jobs, a professional designer is required by local codes. (See Chapter 1.)

Whether you draw the plan or not, every electrician should know how to plan and design an electrical system. Otherwise, how will you recognize a mistake in plans prepared by others? Even if the plans you receive don't have errors or omissions, they won't tell you every detail of the work. A solid background in planning and design will let you figure out a bid without costly mistakes. You'll be

able to spot ways to save time and money without violating codes or the plan requirements.

If you draw up the plan yourself, you may have to submit it to the architect or owner for approval. Any changes you make afterwards should be approved, again by the architect or owner. If you're working from a professionally designed plan, keep in mind that you *cannot* make changes in the plan on your own. You must discuss changes with the designer and give your reasons. If they are permitted, be sure a *change order* is issued. (See Chapter 11.)

The following sections in this chapter will show you how to plan a system that's safe and practical. Study them carefully. Know how to recognize good electrical plans, know how to improve a plan. Be alert to wasteful or unnecessary requirements in the plans you receive. Try to get these changed before you submit your bid and begin work. For your own plans, work carefully and in detail. Follow codes and the client's wishes and you'll have a safe job and a satisfied customer. And money in your pocket.

Working with Plans and Blueprints
Electrical plans generally include the following:
- Wiring Plans
- Schedules
- Riser Diagrams
- Details

Wiring plans are drawn on the architectural plans of the building, usually on the *floor plan*. This is the part of the plan drawn as if viewed from above. An *elevation* is a drawing which shows the building as viewed from the front, back or side.

Wiring plans show the locations of switches, receptacles and fixtures. A typical wiring plan is shown in Figure 5-1. Lines representing the wires are drawn between switches, receptacles and fixtures to show *general* wire location. Numbers and letters show the circuits and fixture types the designer requires. The rooms and appliances are labeled, and the plan itself is identified by a label. Below this label is the *plan scale*. This scale shows the relationship between the drawing and the actual building size.

A *schedule* is a list on a plan which brings together in one convenient location some information needed to understand the plan. A *symbol schedule* (Figure 5-2) identifies the symbols used. A *fixture schedule* (Figure 5-3) shows the symbols used to identify each fixture on the plan. A *panel schedule* provides information about the types and sizes of service equipment.

A *riser diagram* is often used for service equipment (Figure 5-4), intercoms, cable t.v. networks and fire alarm systems (Figure 5-5). It's a simplified drawing, not in any scale, which shows the equipment and terminals needed and gives a general idea of how the equipment is wired. It may also show general locations of individual panels next to one another. But the actual way to mount the panels is left up to the installer, using code rules and common sense.

Many plans include *details*. These are enlarged drawings of a specific area or item that explain what is required for a certain installation. See Figure 5-6. A *section detail* shows a part of a building or piece of equipment as if it's been sliced in half.

Larger jobs will have a book of specifications compiled by the engineer or architect. Spec books include fixture and equipment types, working conditions, installation methods and contractor responsibilities. The specifications in the book are binding on the contractor when he signs to do the work. Changes can't be made without authorization of the architect or project manager. Spec books are very important when bidding and estimating a job, as explained in Chapter 11.

Electrical plans can be simple or complex, depending on the type of building and the designer. When studying plans, you may find that some information is missing. Or perhaps the designer has violated some provision in the code. Be sure to clear up any questions before proceeding with either the bidding or the work.

Preliminary Steps

The first step in planning an electrical system is to get a set of architectural plans. For simple installations, the floor plan alone may be all you need. For large residences and commercial buildings, you'll probably need a complete plan set. Get the plans from the builder, architect or owner.

Before you begin planning a job, have a talk with the architect or owner. Find out about the heating and cooling systems desired and about the major appliances such as hot water heaters, stoves, washers and dryers. Ask about special needs such as a basement workshop.

In a commercial building, task lighting of work areas usually supplements general illumination. Special equipment will need proper circuits and connections. Although you can get some of this information from the room labels on the plan, it's best to check with the architect or owner.

Don't begin your plan until you understand the building to be constructed. During the planning stage, more questions will come up. Make a note on each question so you can take them up with the architect or owner. Complete as much of the planning as possible before requesting more information. This will save both your time and the architect's.

It isn't hard to draw electrical plans. The following pages explain what you'll need to know. Use the plans in Figures 5-7 through 5-9 for practice. Study the symbol schedule in Figure 5-2, and practice drawing the different symbols.

The pencils, scale rulers and templates you'll need are available at drafting supply stores. The floor plan should be fastened with masking tape to a drawing board, but any smooth surface works nearly as well.

On remodeling jobs you may have to draw the floor plans from scratch. Use drafting paper with a non-reproducing blue grid. The grid lines make it easier to draw a straight line the right length. To reproduce the building, you'll need field measurements and a rough field sketch. The sketch can then be re-drawn on the grid paper with a scale ruler, using any suitable scale. As a rule, the larger the building, the smaller the scale. But a scale too small means a plan that's cluttered and hard to read.

The order to follow when drawing the plan is your choice. But many electrical designers do all the lighting, then the receptacles, then the service. Be sure and read the Lighting and Receptacle Requirements sections in this chapter first so you follow the code.

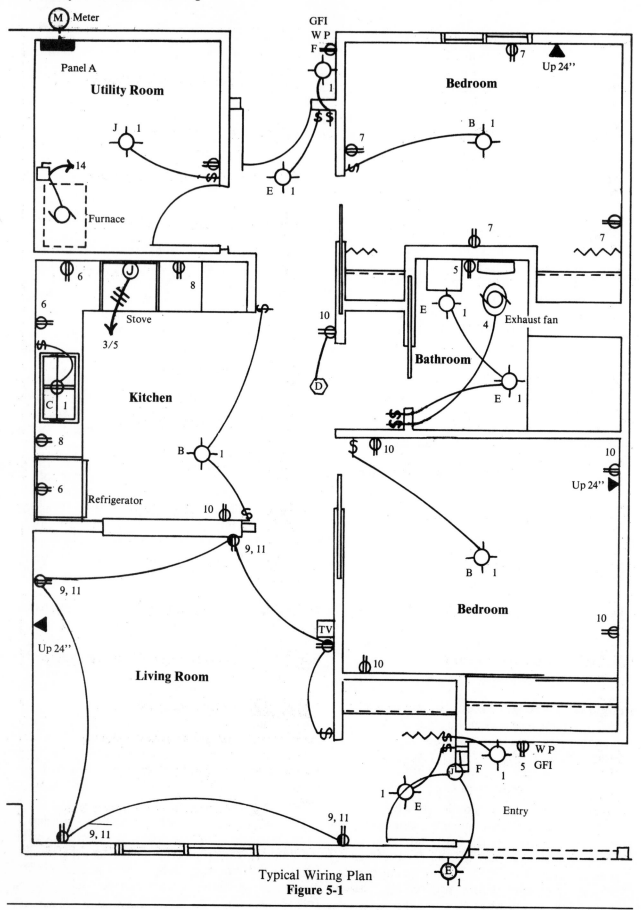

Typical Wiring Plan
Figure 5-1

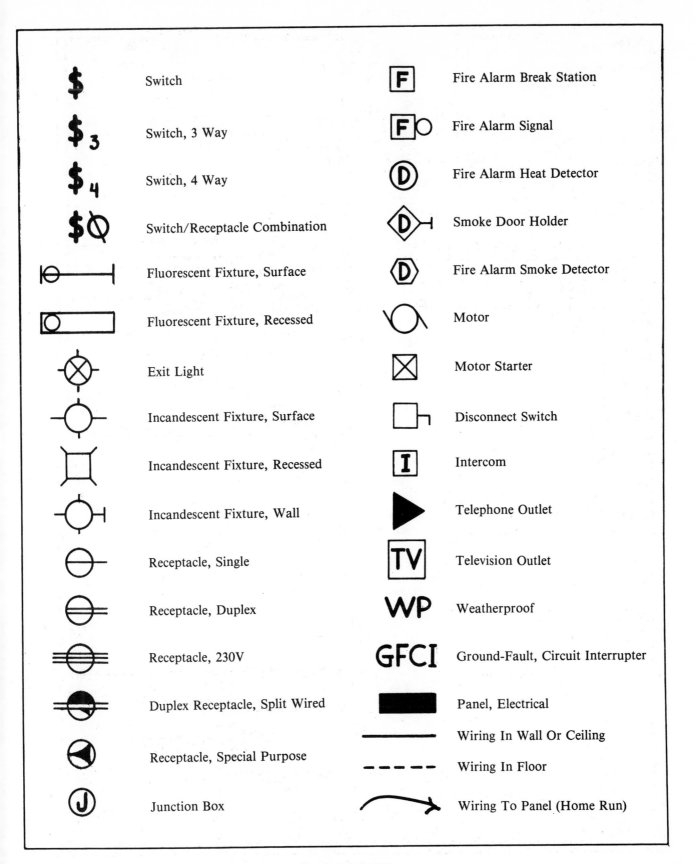

Symbol Schedule
Figure 5-2

Fixture Schedule

Symbol	Manufacturer	Catalog Number	Lamps	Mounting	Trim	Remarks
A	Prestolite	9402	2-75W	Surface	White	
B	Prestolite	9400	1-75W	Surface	White	
C	Prestolite	8362-LPF	1-20W	Surface	White	
D	Prestolite	9453	1-75W	Surface	Black	
E	Prestolite	488HS-1	1-100W	Recess	White	
F	G-E	IC-260	2-60W	Surface	None	
G	Spaulding	803-100-120	1-100WMV	Post	White	
H	Prestolite	WB-24	1-100W	Wall	Aluminum	
K	Metalux	M-240A	2-F40T12	Surface	White	

Fixture Schedule
Figure 5-3

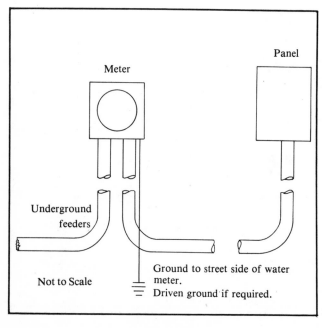

Service Riser Diagram
Figure 5-4

If you start with the lighting, draw in the light fixtures for each room. Lighting fixtures are generally located in the center of the ceiling. A quick way to find the center of a square or rectangular room is to lay a ruler diagonally on the plans, from corner to corner. Make a light, short line in the center. Do it again, using the opposite corners. Where the lines cross is the room center.

To find locations in odd-shaped rooms, measure with a scale ruler that matches the plan scale. However, be careful when scaling plans. Architects sometimes give room dimensions in feet and inches that don't match what the plan shows. If the plan scale doesn't agree with the architect's dimensions, check with him to see which is correct.

After the lighting is drawn, locate the wall switches. Put the switches as close as practical to the door openings. Be sure the direction of the door swing is indicated on the plan so the switch doesn't end up behind the door. In locations such as large rooms, stairs and hallways, three-way switches are appropriate.

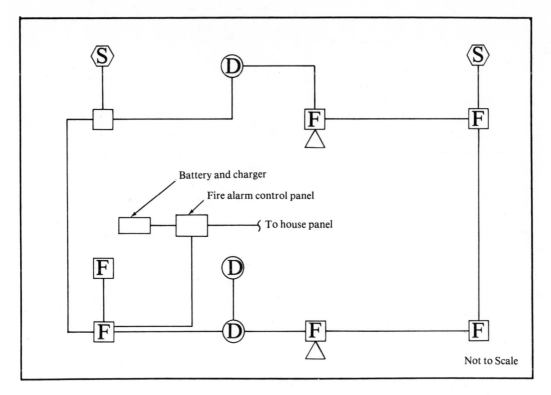

Fire Alarm Riser Diagram
Figure 5-5

Now, locate all the convenience receptacles and equipment connections. Go back and connect the switches to the light fixtures or appliances that they control. Put in circuit numbers and fixture identification (Chapter 6) as soon as they're determined.

Place labels in the space that's left. The lettering should be neatly printed. Lettering guides allow you to draw neat, clear letters. Keep the lettering parallel to the structure lines. All lettering should be done so it can be read without turning the plan.

Planning for Single Dwellings
This section offers recommendations for locating lighting outlets, receptacles and equipment connections in a typical home. It covers the basic NEC requirements. Of course, every home is different. Some homes will need more detailed planning, depending on how much the owner is willing to spend.

Lighting Requirements
The NEC requires at least one wall-switched lighting fixture for each habitable room (Section 210-C). The NEC defines a habitable room as a "kitchen, family room, dining room, living room,

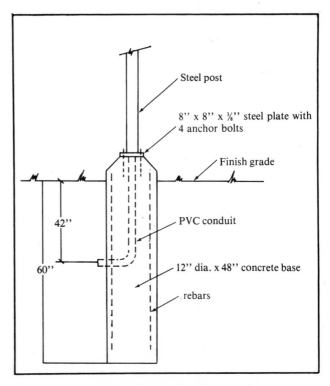

Lamp Post Detail
Figure 5-6

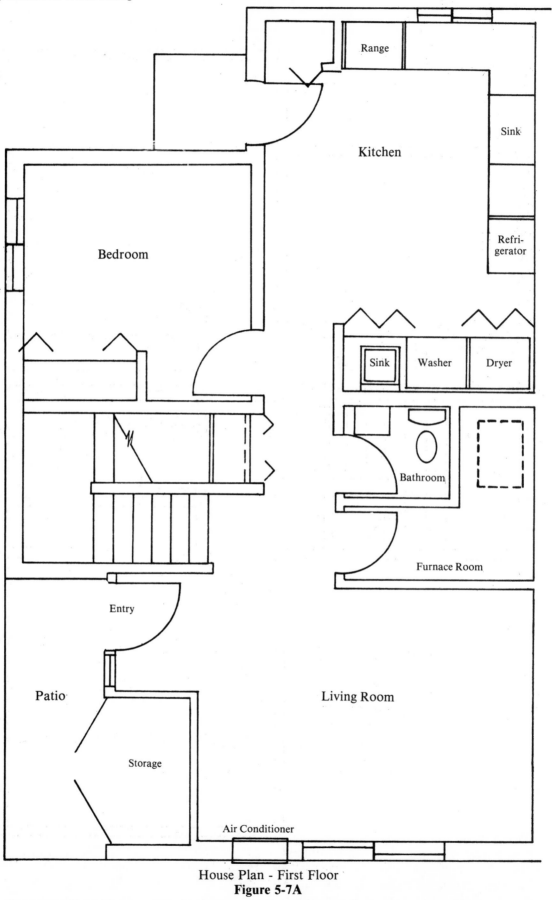

House Plan - First Floor
Figure 5-7A

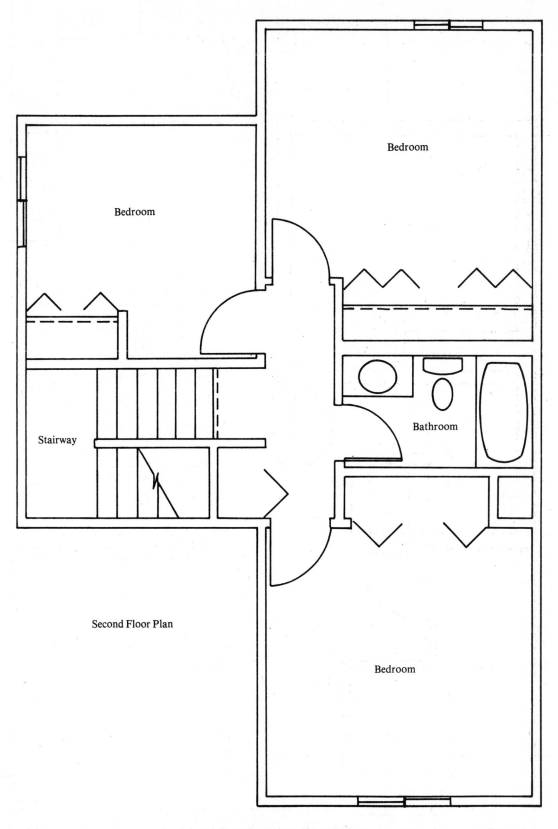

Bedroom

Bedroom

Stairway

Bathroom

Second Floor Plan

Bedroom

House Plan - Second Floor
Figure 5-7B

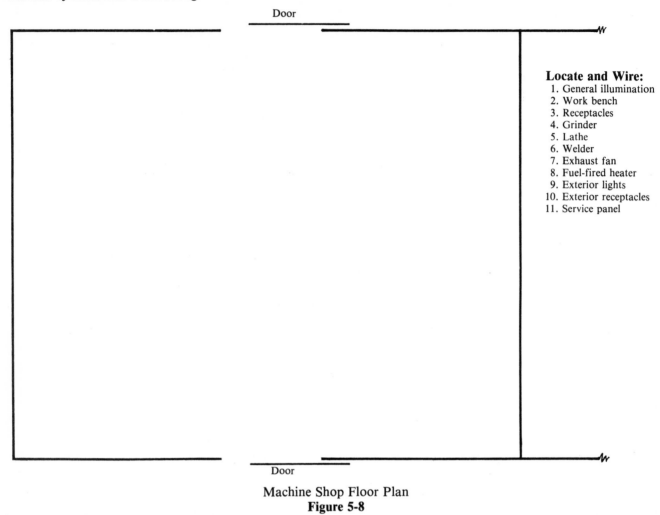

Machine Shop Floor Plan
Figure 5-8

Locate and Wire:
1. General illumination
2. Work bench
3. Receptacles
4. Grinder
5. Lathe
6. Welder
7. Exhaust fan
8. Fuel-fired heater
9. Exterior lights
10. Exterior receptacles
11. Service panel

parlor, library, den, sun room, bedroom, recreation room or similar rooms." The fixture can be either wall or ceiling mounted. This method is generally followed for all rooms except the living room. While acceptable for living rooms, the common practice is to connect wall switches (see Chapter 7) to the top half of the living room convenience outlets. (Convenience outlets are the duplex receptacles located 12 feet apart on the walls of a room. They're usually mounted 12 inches off the floor.)

Wall-switched receptacles operate lamps and other portable lighting fixtures. The fixtures can be changed or moved freely. While the NEC permits wall-switched receptacles for lighting in all rooms except the kitchen, they're most practical in the living room.

Convenience Receptacle Requirements
The NEC also regulates the minimum number of receptacles in habitable rooms (Section 210-C).

The general NEC rule requires convenience outlets in rooms so that no point along the floor line of the wall is more than six feet from an outlet. In other words, outlets are spaced 12 feet apart. Any wall more than two feet long must meet this rule.

Partial walls or permanent room dividers are considered to be walls. Wall space behind open doors is also counted. Outlets that are part of an appliance or lighting fixture don't count towards meeting the spacing requirement.

Certain areas in the kitchen and bathroom have added requirements. These will be pointed out in following sections.

The general practice, unless indicated by code or common sense, is to mount convenience outlets 12 inches from the floor. Wall switches are mounted 44 inches off the floor.

Kitchens
A kitchen will need a ceiling light and one over the sink. Each should be controlled by a wall switch. In

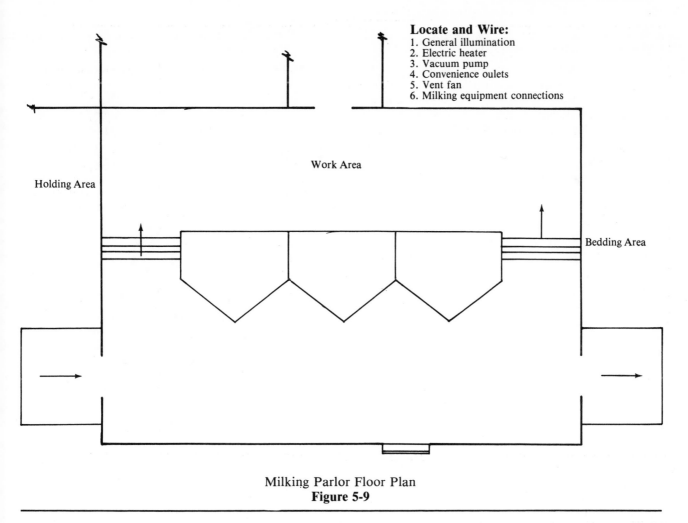

Locate and Wire:
1. General illumination
2. Electric heater
3. Vacuum pump
4. Convenience oulets
5. Vent fan
6. Milking equipment connections

Work Area

Holding Area

Bedding Area

Milking Parlor Floor Plan
Figure 5-9

a small kitchen with no dining area, the sink lighting outlet alone may fulfill code requirements. If the kitchen is large, with more than one doorway, three-way switching may be needed for the ceiling light.

Extra lighting and an exhaust fan should be located over the stove. Some kitchens have a nook for a desk; this will need lighting and perhaps a telephone outlet. Some kitchens have decorative hanging light fixtures mounted over the kitchen table. Position these carefully, so they won't be in the way. If the table is moved, the fixture could become an obstacle.

Be sure to provide plenty of receptacles in the kitchen. The NEC requires at least one outlet for each separate counter space wider than 12 inches. Counter tops divided by ranges, sinks and refrigerators are considered separate areas. Outlets should be located every 4 feet along the counter top and 8 inches above the counter top. If you're putting in outlets before the counter has been installed, place the outlet 44 inches from the floor.

Use split-receptacle outlets for counters and work spaces. These outlets have separately circuited top and bottom receptacles to prevent overloads (See Chapter 7). To avoid a confusion of switches and outlets along the counter area, use dual receptacles and receptacle-switch combinations.

The refrigerator must have an outlet. And if there's a separate freezer in the kitchen, it should also have its own outlet. Mount these outlets 44" off the floor.

Gas stoves will require a 115-volt outlet if they have lights and a clock; an electric stove will use a 115-230-volt circuit. A dishwasher and a waste disposal will each need a proper size circuit and switching. These large kitchen appliances should be circuited separately. Check your local codes.

Convenience outlets should be located along open wall space as required. Outlets located behind appliances don't count toward the NEC outlet requirement. If the floor plan shows a table location, one outlet can be placed 36 inches above the floor next to it. Built-in ironing boards and wall clocks will need specially placed outlets.

The NEC requires that *all dwellings have a minimum of two 20-amp circuits in the kitchen.* These circuits are used for small appliance loads, convenience outlets, counter outlets and refrigeration equipment. The circuits must be located in the kitchen and dining areas. They can't serve outlets in other areas. Small appliance circuit design is explained in Chapter 6 in the section on branch circuits.

Bedrooms

General lighting in the bedroom can be from a wall-switched ceiling fixture. More lighting may be placed above mirrors, dressing tables and at the bedside. Plan for ceiling or wall mounted fixtures in these areas if the owner knows exactly what room arrangement he will use. But permanent fixtures may prevent or hinder rearranging room furniture. It's usually better to install plenty of convenience outlets for portable lighting fixtures.

The general NEC rule on spacing outlets along a wall applies to bedrooms: every point must be within six feet of a receptacle.

Often the designer will place the first outlet six feet from the door, with additional outlets every 12 feet. Although this follows the code, it may be poor practice in the average-size bedroom. An outlet will end up in the center of a wall where it's hidden by the bed headboard. Plugs and cords become tangled and squeezed by the headboard. It's better to place outlets three to four feet on either side of the center of the wall. Of course, this increases the number of outlets and also the cost of the job. But it's a small price to pay for 30 years or more of convenience.

Include a proper-size equipment outlet for a window- or wall-mounted room air conditioner if central air conditioning isn't in the plan.

Bathrooms

Bathrooms should have a wall-switched ceiling light. Rather than locate this fixture in the center of the ceiling, put it nearer to the bathtub. If the bathroom isn't too large, the main light source may be wall-mounted above the mirror. A more detailed lighting design might include a ceiling fixture, decorative fixtures at the mirror and a light fixture over the tub.

Fixtures mounted by a mirror should distribute light evenly over the entire mirror. Avoid having a single light source on one side of the mirror. Balance it with a fixture on the other side. And don't place the overhead fixture between the user

and the mirror; it should be directly over the mirror to prevent shadows.

Bathrooms with shower enclosures should have a moisture-proof light fixture over the shower. Mount the switch outside the enclosure.

The NEC requires one convenience outlet next to the wash basin, 44 inches off the floor. The outlet must be a ground-fault, circuit-interrupter type. You may need two outlets for a large washbasin area. These should be the only outlets in the bathroom.

Bathroom Exhaust Fan
Figure 5-10

Most bathrooms have exhaust fans. See Figure 5-10. In bathrooms without windows, fans are needed to remove moisture and odors.

Exhaust fans may be separately switched or switched with the light fixture. Combination fixtures are also available. These have a heating element or lamp, with a fan that circulates the hot air. The light fixture may also be a part of the combination. As a rule, the different functions of the multi-purpose fixture should have separate switches.

Check local codes for exhaust fan and multi-purpose fixture installation and switching requirements.

Living and Dining Rooms

These rooms may have a ceiling light fixture with a wall switch. Or, as mentioned earlier in this chapter, the top half of the convenience outlets may be wall-switched. Location of convenience outlets must conform to the general NEC rule: no

Courtesy: Thomas Industries

Track Lighting
Figure 5-11

point along the wall can be more than 6 feet from an outlet.

Areas such as reading nooks, mantles, fireplaces, conversation pits and home entertainment centers will need outlets for lighting and appliances. Track lighting is a good way to get both decorative and work task lighting from a single outlet (Figures 5-11 and 5-12). Built-in stereos, TV's, and fireplace heat circulators will need outlets.

A formal dining area will have a buffet or sideboard. Outlets should be installed for food warmers and other appliances. Many owners like to have a dimmer control switch on lighting fixtures above decorative furnishings or special room areas.

Closets, Halls and Stairways

Closets more than two feet deep should have their own light fixture. These fixtures should be wall-switched rather than pull-chain operated. Pull-chains are hard to find in the dark. Closet fixtures can be mounted on either a wall or the ceiling. Allow 18 inches between the fixture and any shelves where combustible materials are likely to be stored.

Large closets may need a receptacle. You may want to wire the receptacle to the wall switch.

Halls and stairways must have wall-switched lighting fixtures. A hall that's long or irregular in shape needs more than one light. On a stairway, put the lighting fixture where it will illuminate all

Courtesy: Thomas Industries

Track Lighting
Figure 5-12

flights well. Avoid putting fixtures where it will be difficult to replace the lamp.

Three-way switching is a good idea in a hallway. For a stairway between habitable floors it's a necessity. The switches should be near the head and foot of the stairs but far enough away so the user won't trip or fall when reaching for the switch.

For hallways, convenience outlets should be placed so that no point along the hall is more than 15 feet from an outlet. Any hall with a floor area over 25 square feet should have an outlet. If no outlet is close to the foyer or entryway, put one there.

Utility Rooms, Basements and Laundry Rooms
Utility rooms are for the building's mechanical

systems. They may also house the laundry equipment. The NEC requires at least one lighting outlet for these rooms. The outlet doesn't have to be wall-switched.

The mechanical equipment will need separate circuits for equipment connections. Central air conditioning units, electric hot-water heaters and clothes dryers need 230-volt circuits. The washer will need a 115-volt circuit. A fuel-fired furnace uses 115-volt power. Check the appliance manufacturer's specifications to find wattage and other information. Circuiting is covered in Chapter Six.

Laundry equipment is sometimes placed in a separate room. The work area should have good lighting. Fixtures should be wall-switched. A convenience receptacle, in addition to the appliance receptacles, is essential. An exhaust fan is desirable if the laundry area is enclosed and doesn't have a window.

Whether the laundry equipment is located in a separate room, the utility room or the basement, *all outlets in the laundry area must be on a separate 20-amp circuit, with no other outlets.* Outlets for the appliances must be within six feet of the area set aside for the appliance. Outlets should be mounted 44 inches off the floor.

For basements, the NEC requires a minimum of one lighting outlet. Most residential basements have a light fixture near the foot of the stairs controlled by a wall-switch at the head of the stairs. Additional lighting could include lampholders or fixtures controlled either by the same switch or by pull-chains.

In basements with incandescent lamps, a general rule of thumb is that 0.3 of a watt is needed for each one square foot. In other words, a 100-watt bulb will light about 300 square feet. This is good enough for general lighting, but work areas should have additional light.

Besides outlets for mechanical and laundry equipment as described above, the NEC requires at least one convenience outlet in the basement.

Freezers, well systems and sump pumps are often placed in the basement. This equipment needs separate circuits and outlets to keep an overload from tripping the breaker. No one wants to be surprised by a freezer full of thawed food or a flooded basement.

Consider the future wiring needs of the homeowners. Many basements are converted into a recreation room or workshop. Provide enough capacity to meet these requirements. It costs far less to add capacity when the home is built than to rewire it later.

Attics, Enclosed Porches and Garages
Attics which are to be finished rooms at a future date should be wired with that in mind. If it's left unfinished but has a permanent stairway, a wall-switched light fixture must be installed at the foot of the stairs. A convenience outlet should also be provided. If the attic has a folding stair or ladder, a pull-chain light fixture is all you need above the attic opening.

As a general rule, an enclosed porch or breezeway must have both a wall-switched light fixture and a convenience outlet. Since it may be hard to decide what is enclosed and what isn't, check with your local inspector. Sometimes the minimum square footage of an area is used to determine the requirements.

If the porch or breezeway is open to the weather, weatherproof outlets are required. If the enclosure connects the house with the garage, use three-way controls for the light fixture.

An attached garage must have a wall-switched lighting outlet and one GFCI convenience outlet. Mount the GFCI outlet between 44 inches and 66 inches off the floor. GFCI outlets aren't necessary for garage door openers, freezers or other appliances. The outlets for these appliances, however, are located where they aren't accessible for general use; they won't fulfill the general convenience outlet requirement. Additional convenience outlets can be installed as needed. They should be the GFCI type.

Exterior Entrances and Patios
The NEC requires at least one outdoor convenience receptacle for one- and two-family dwellings. A large dwelling may need more than one. Place the receptacle near a door or the patio, at least 18 inches above the grade. Use a weatherproof GFCI receptacle. The outlet can be controlled by a wall-switch inside the dwelling, if desired. Outlets that are parts of lighting fixtures do not meet the exterior outlet requirement.

Lighting outlets are required at exterior entrances. Vehicle doors on attached garages are not considered exterior entrances. Exterior lights should be controlled by switches inside the house.

Here are two important considerations that aren't in any code book but should be obvious to everyone who designs electrical systems in homes:

1. Provide good lighting on stairs and sidewalk leading to the entrance.

2. Provide lighting at entrances so building occupants can see who is at their door.

If the sidewalk is long, post-mounted lighting is a nice feature. If the dwelling has several exterior lights, control them automatically with a photoelectric switch or a timer. You may even suggest a master control switch for exterior lights in the master bedroom.

Planning Commercial and Multi-family Dwellings
The same basic ideas used in planning single-family dwellings are used in commercial and multiple housing design. Types and uses of rooms and work areas dictate the kind of lighting to be installed, the number of convenience outlets, and mechanical equipment connections.

There are, however, several points that deserve special attention.

Code Restrictions

Because most commercial buildings are open to public use, national, state, and local codes are more strict.

A bathroom exhaust fan may be desirable in a residence. It's generally required in a public restroom. Buildings with low light levels such as cocktail lounges and restaurants will need local lighting on stairs between floor levels. Electrical work in a workshop or commercial kitchen is tightly regulated.

Buildings that house many people, such as apartment complexes, hotels, and theaters, must have emergency lighting and exit signs. Fire alarm systems are also required.

Many commercial systems use high voltage power. Generally, as the voltage of the system increases, NEC regulations become more strict. Wiring regulations are also more strict if the building is used for manufacture or storage of materials which may explode or burn.

Start planning for commercial or multiple-housing buildings the same way as for a single-family dwelling. Other items to consider are detailed in the rest of this chapter. Once the plans are drawn, they must be submitted to the state or local inspector for approval. Without approval, no permit can be issued.

Multiple Housing

Multiple dwellings is the classification which includes duplexes, townhouses, condominiums and similar residences. Essentially, these are private dwellings that share a common wall and roof. They can be treated as private dwellings for design and code purposes. But the NEC has special rules for service equipment and installation in these buildings. The rules are explained in Chapter 6.

Units which open out into a common hall or entrance are grouped into a class called *multiple-occupancy* buildings. Typically, an apartment building is multiple occupancy. Again, each unit can be designed as a private dwelling. But because the building has common areas open to the public, additional rules apply. State and local inspectors are the best source of information when designing a multiple-occupancy building. They should also be consulted when the building has a special use such as housing for the elderly or nursing home care.

Emergency Systems

The National Fire Protection Association has established standards for emergency lighting and power needed for safe exiting and panic control.

Almost every building occupied by a large number of people, either temporarily or permanently, now needs some kind of emergency system. Auditoriums, hotels, theaters, apartment buildings and health care centers are common examples. Other buildings may need emergency systems as determined by the agency having authority. Usually this would be the state and local building inspector.

Emergency circuits generally serve fire exit signs at exterior doors, fire alarm systems and emergency lights in halls and stairs. These circuits must be wired independent of non-emergency circuits. In some cases they can't share the same raceway with other circuits.

Emergency circuits must have an independent power source. In some areas and for some circuits, you meet this requirements by providing a separate service ahead of the main service disconnect. Other circuits use storage batteries or generators. Switches and other service equipment for emergency systems must either be located in areas where access is limited to authorized people or have locked panel covers and switch boxes.

Talk to your inspector when planning an emergency electrical system. Manufacturers are also good sources of information about equipment and installation. NEC requirements are detailed in Articles 700 and 760 of the code.

Intercommunication Systems

Intercom systems (Figure 5-13) range from a simple one-way call box to complicated systems with music, door-control, and multiple calling. Most manufacturers supply—along with the equipment—a schematic plan, installation instructions, and color coded wiring.

Intercom wires are not permitted in raceways with any electrical wires, unless separated by a partition. In open wiring systems and shaft runs, the intercom wires must be kept 2 inches apart from electrical conductors. Some exceptions are permitted, most often if the electrical conductor is enclosed in a raceway or has suitable insulation. Article 800 of the NEC details intercom system rules.

Clearance between intercom and electrical wires provides a margin of safety if there is a short circuit. It also prevents static caused by interference in the intercom system. Intercom wires should also be kept away from telephone lines for this reason.

When intercom lines are located outside, plan for enough clearance between roofs and other conductors (Article 800, NEC). The lines must be attached to poles and buildings as described in the code. Protective devices such as arrestors must be

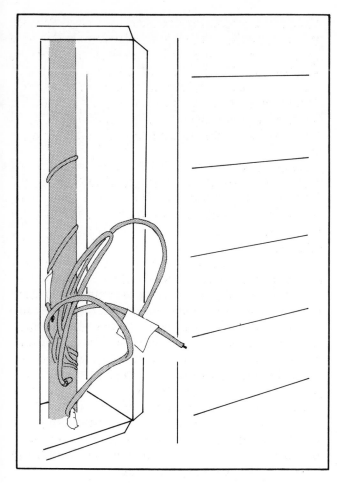

Roughed-In Intercom Outlet
Figure 5-13

used where the intercom wires enter the building. These provide protection in case the wire is accidentally charged with high voltage. Ground the arrestor and other intercom equipment to the system ground.

The amplifier for the intercom system is usually located close to the master station. It will require a 115-volt outlet.

Television Systems

Many multiple dwellings have a master antenna television system (MATV) to bring in a strong signal for the sets in the building. Sometimes this system is designed to hook up with a community or cable company system. Otherwise, it may end in a large, roof-top antenna. MATV installation is generally included in the electrical contract. You may want to subcontract this work to a MATV system specialist.

A basic MATV system includes an antenna, a power amplifier that boosts the signal, and coaxial cable to carry the signal. The system will also use splitters for branch lines and tap-off units, to compensate for loss of signal. Most manufacturers will design a system for a building if you supply a set of plans.

Rooftop antennas can't be mounted on service masts or poles carrying electrical wires. They must be strong enough to carry snow and ice loads and be kept well clear of power lines to prevent accidental contact. The antenna and its cable must be grounded as close as practical to where it enters the building. The cable has to run at least 8 feet above the roof, if the roof has an opening for pedestrian traffic. It must be installed to prevent accidental contact with electrical wires and be equipped with suitable protectors.

Inside the building, MATV cable follows the same rules as intercom wire. Where the cable ends in an outlet, the connection box should be mounted 12 inches above the floor. Details on MATV antenna and cable installations are in NEC Articles 810 and 820.

Exterior Lighting and Receptacles

Exterior lighting for commercial and multiple dwellings serves three purposes:

1. It decorates and illuminates buildings and signs identifying them.

2. It safeguards the premises against illegal entry.

3. It provides a safe entrance and exit for people using the building at night.

For some jobs, one lighting fixture can serve all three functions.

Multiple dwelling entrances should be lit the same way as single-family dwelling entrances. A weatherproof, GFCI receptacle is also needed. For multiple-occupancy buildings, all entrances and exits must have a light fixture. They may be controlled by separate wall switches. Usually, it's better to have them switched on and off by a time clock or photoelectric cell.

Large buildings and parking areas should have post-mounted lights. Your local building code will spell out what's required. Post fixtures placed in parking areas should be mounted on concrete bases 18 inches above the grade or protected by steel guard posts.

Shorter posts that hold fixtures in lawn areas should be installed as directed by the manufacturer. If the light is to be on more than three hours a night, it's best to use a mercury, sodium vapor, or metal halide lamp for general lighting. These lights are more efficient and create few shadows. But the color of the light given off by these fixtures is different from that of incandescent lamps.

Lighting that decorates or highlights the building should use incandescent spotlight-type fixtures. Use high-intensity fixtures (mercury/sodium/halide) in parking and lawn areas where light color is not as important.

Service and maintenance doorways should have a light fixture and a weatherproof GFCI receptacle. If the building is large, more exterior receptacles should be located at 30 to 50 foot intervals.

Signs are covered by more than one code. As a rule, local codes control the size, type, and locations of signs and their mountings. NEC Article 600 covers wiring and the general sturdiness of the signs.

Exterior lighting will often call for buried cable or wiring that runs along the building exterior. These are explained in Chapter 7.

Planning for Farm Buildings

Most wiring on a farm is about the same as other electrical work. But parts of the job take special knowledge and a unique understanding of the owner's requirements.

Many farms had the original electric service wired to the house or a yard pole. From this point feeder lines are run to the barn and other outbuildings. As the use of electricity on the farm increased, other lines were added. Some or all of this work was probably done by amateur rather than professional electricians. Eventually, this jerry-built system has to be overhauled.

If you handle farm work, you'll usually be replacing or remodeling the existing distribution system rather than doing strictly new construction.

Even the way you handle a farm job can be different. New construction is usually done from beginning to end before the owner takes occupancy. The farmer or rancher who hires you may want an electrical plan that can be done in phases over several years as the need arises and as money is available. You'll find that farmstead wiring can be done very easily on a step-by-step basis.

Planning for future needs is also an important part of farm wiring. Labor-saving machinery has increased productivity on the farm. Much of this machinery is electrically operated. More of it will be available in the future. This makes it almost impossible to "overwire" a farm. Correcting errors later is always more expensive than doing a good job the first time. Planning and installing for future electrical loads is never more important than in farm work.

Code Restrictions

Most agricultural buildings are exposed to dust, water and corrosive substances. NEC farm requirements deal with these problems in Article 547. Wiring, fittings, boxes, flexible connections, switches, and lighting fixtures must be weatherproofed or designed for protection against dust and dirt. Flexible connections and guards must be used where the wiring may undergo physical abuse. All parts of the system must be properly grounded.

Residences on farms are covered by dwelling code requirements. Underground and exterior wiring, often used on farmsteads, are covered by the sections on those subjects in the NEC.

Be sure to check with the local power company before beginning work. They'll have requirements of their own.

Drawing Up a Plan

Start a farm job by following the general planning steps that are outlined at the beginning of this chapter. In addition to floor plans, you'll need a sketch showing the location and use of each building. This can be a rough sketch, but it should show the location of the yard pole, if there is one. Mark the distance between buildings as accurately as possible. The sketch should also note the service and feeder lines and sizes to each building. See Figure 5-14.

After sketching the plot plan, draw floor plans of the buildings where the work will be done. Draw in existing wiring, if there is any. Mark changes and removals, as well as the new work, on this floor plan.

When the work is completed, this plan becomes a permanent record of the wiring system and how you changed it.

Farm Residences

Farmhouse wiring is like wiring a single-family dwelling. Follow the steps outlined earlier in this chapter. The remodeling section in Chapter 8 may also be useful.

Equipment Connections

All farm buildings need equipment connections. Barns will have ventilating fans and feed-handling equipment. Brooder and maternity pens will often need heat lamps. Milkhouses and parlors use water heaters and vacuum pumps. Workshops will need suitable wiring for welders and similar equipment. It isn't unusual to find as many as fifty different equipment connections on a modern farm.

Plan these connections to provide adequate voltage and amperage. Circuit size will depend on current, voltage and distance requirements.

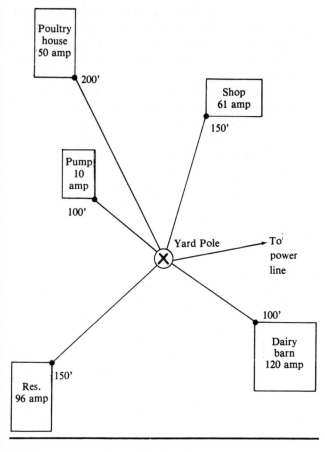

Typical Farm Sketch
Figure 5-14

the lighting outlets along the center-line of the litter alley. The outlets should be spaced 10 to 15 feet apart. A good rule of thumb is to have an outlet at every other stall divider. If fluorescent lighting is used, a continuous row of two-lamp, standard white fluorescents in semi-direct fixtures should be installed along the alley center-line.

For a face-in arrangement, lighting outlets are placed 12 inches back from the gutter over the alley, directly behind every other stall divider. A single-lamp continuous row of fluorescent fixtures may also be mounted here if fluorescent lighting is preferred.

Light the feed alley with lighting outlets located along the center-line about 12 feet apart or about 1.5 times the fixture mounting height.

Feed and litter alley lights should be wall-switched. Litter alley lights should be controlled by three-way switches from either end of the alley. Switch feed alley lights from a central location on the litter alley. Mount the switches at least 44 inches off the floor. Switches which control outdoor and haymow lights need a pilot light to show when the light is on.

Convenience receptacles—20 amp—should be located every 20 feet along the litter alley. Put them 6 feet above the floor to avoid damage by animals. Where cows face in, the receptacles can be mounted on the outside walls. Where cows face out, use post or beam mounted receptacles. A few receptacles are handy when spaced along the feed alley. Receptacles mounted next to the light switches located at each end of the alley are also convenient.

Provide equipment connections for ventilating fans, gutter cleaners, and similar equipment.

Loose and Free-stall Housing
Lighting outlets should be placed where they make chores easier. The fixtures should light all corners so there are no long shadows. In open areas, plan the fixture spacing at 1.5 to 3 times their mounting height. When lighting feed alleys for feed troughs, the spacing should be 1.5 to 2 times the mounting height.

Lamps that are mounted at heights of 12 feet or more and can be expected to be on for 3 or more hours at a time should be mercury vapor, metal halide, or high-pressure sodium lamps. These luminaires give good light while using little power. Incandescent lamps work best for frequent on-off switching.

Each group of lights needs a separate wall switch. Consider switching a single light in a central location separately. This provides light so a person

As a general rule, any equipment drawing 30 amps or more should be on a separate branch circuit. This is also true for any motor over 1/2 hp. It's possible to put the same types of equipment on the same circuit—for example, three small workshop motors or two barn ventilating fans—as long as the circuits are large enough. However, it's best not to have unlike equipment connected to the same circuit, especially if one piece of equipment is needed for protection of livestock. Common sense will tell you which equipment needs a separate circuit, even if it draws low amperage.

Size and location of equipment depends on economy and convenience. Your electrical utility, state agriculture agent, or the equipment manufacturer will have valuable suggestions. Manufacturers also furnish wiring information. Circuit calculations and installation data for motors and other equipment are in Chapter Six.

Stanchion Barns
To light a face-out stanchion arrangement, place

can pass through to another section of the building.

Plan a convenience receptacle next to each major entrance and beside places where portable clippers, fans, and other equipment will be used. Place them 6 feet above the floor.

Equipment connections will be needed for ventilating fans, water tank heating, feed, and gutter equipment.

Box Stalls and Pens

One ceiling-mounted light outlet is needed in each bull, maternity, or calf pen. It should be wall-switched outside the pen.

Convenience receptacles should be placed in stalls and pens as needed. When pens have low partitions, one outlet can serve two or more pens.

Calf Housing

Provide one lighting outlet for each 50 to 70 square feet of floor space in calf stall areas. Outlets can be wall-switched as a group. In stall areas, one outlet is needed above each stall. Switches are always located outside the stall.

In stall areas, put a convenience receptacle on each outside wall. In pen areas, it may be most convenient to have a receptacle next to each light fixture. Wire it to stay hot when the light is switched off. As a rule of thumb, any receptacle should be within 25 feet of the place where it's likely to be needed.

Calves don't give off enough body heat to maintain proper housing temperatures during cold months. In a well-insulated calf building, the amount of heat needed can be calculated from the following:

$$3 \text{ Watts per square foot of glass}$$
$$+$$
$$1\frac{1}{2} \text{ Watts per square foot of wall}$$
$$+$$
$$1 \text{ Watt per pound of calf.}$$

A forced-air, ceiling, or wall-mounted space heater works best. Although radiant heaters can be used, the heat can't be directed full-force on the animals. Space heaters and ventilating fans should be on separate circuits.

Milking Parlors

Milking parlors have many different stall arrangements. Common sense is probably your best guide here. A good lighting plan provides one lighting outlet for each 36 square feet of work area.

That means about 2.5 outlets per stall. If fluorescent lights are used, the lamps should be in a continuous row over 80% of the room length. Whether incandescent or fluorescent, allow one row of lamps for the operator and one row for *each* row of stalls. Locate one lighting fixture outside each animal entrance. Weatherproof wall-switches should be used for all lighting.

Where incandescent lighting is used behind the cows (other than in herringbone parlors), lighting outlets should be placed about one foot from the outer wall. Swivel-type fixtures work best because each lamp can be adjusted to the right position.

Plan for a 20-amp, weatherproof convenience receptacle at each end of the operator's work area. Put another weatherproof outlet for a waste pump under the pit floor. It's easy to forget equipment connections for fans and pumps. Go over the plan once more before you finish and look just for fan and pump connections you may have missed.

Heating equipment may be needed in the operator's area. One common heating system is a 3500-W quartz tube lamp located along the center line of the operator's area. One lamp each ten feet will give enough heat in most climates. If infra-red heating lamps are used, 50 watts of heating capacity are needed for each square foot of work area. Heat lamps and wiring should be installed on separate circuits. Two or more units may need a three-wire, 120/240-V circuit to prevent overloading. Be sure to follow the manufacturer's recommendations and local codes for wiring and mounting heights.

Milk Houses

Lighting outlets should be over work areas in the milk house. Vapor-tight fixtures are recommended for use over wash-up areas. For the bulk milk tank, you'll need two 150-W PAR lamps in swivel fixtures on the ceiling. They can't be located directly over the tank. Switch milk tank lights separately from the other work lights. A separately switched outdoor light is needed where the tank-truck picks up the milk.

Plan for enough convenience receptacles where they are likely to be needed. A GFCI receptacle is essential next to the sink. Mount it 44 inches off the floor. Other receptacles must be either placed high enough to avoid splashing or be weatherproof.

Equipment connections will be needed for hot water heaters, milk coolers, and ventilation. A small space heater may also be a good idea. Put an outlet for the milk-truck pump on the outside wall. Check with the driver or the dairy representative for details of this mounting.

Poultry Buildings
In a laying house, lighting is often used to increase egg production. How much light and the type depends on the kind of poultry and the geographic location of the building. State agricultural agencies and rural power suppliers can usually recommend the best laying house lighting for their area.

It's always better to provide a few too many lighting outlets rather than not enough. This gives the operator more flexibility in varying the lighting applied.

Keeping this in mind, two calculations can be used.

$$1.\ \text{Maximum No. of Lamps} = \frac{\text{Floor Area (Sq. Ft.)}}{140}$$

$$2.\ \text{Minimum No. of Lamps} = \frac{\text{Floor Area (Sq. Ft.)}}{200}$$

For caged layers, lighting outlets should be spaced 10 feet apart above the aisles and not over 6 feet above the lowest cages. For a general system, the lighting outlets should be spaced from 8 to 12 feet on center along the width of the building and from 10 to 16 feet along the length. The end outlets should be 4 to 5 feet from the walls. Mount fixtures 7 feet above the floor.

There are several ways to control the lights. Manual control is always needed so the operator can turn the lights on and off at any time. Automatic controls will be part of most systems. A time switch can add a period of artificial light to the normal daylight hours. It should be adjustable from 1 to 24 hours. Some types of lighting programs increase or decrease the lighting gradually. These programs need groups of lights circuited separately with separate automatic controls.

Convenience receptacles are needed for general use and for water heating. One receptacle should be installed for every 400 square feet of floor area. Put at least one outlet in every pen at least 3 feet above the floor on walls or posts. If necessary, suspend an outlet from the ceiling on a pendant cord. Use No. 12 stranded cord or larger. Cord types such as SJ, SJT, ST, or S are required. Equipment receptacles should be located as needed.

Brooder houses may be part of the laying house or a separate building. Lighting and receptacle locations are the same as in a laying house. Note that receptacles in the brooder area may be needed to supply heaters. In this case, three-wire, 120/240 volt circuits are recommended.

In a broiler house, plan for one lighting outlet for each 50 square feet of floor area in the records, feed, and control rooms. Broiler growing areas need one outlet in each 400 square feet of floor, or one outlet in each pen. Control is by wall switches and time clocks.

Convenience and equipment receptacles should be installed as outlined for laying houses.

Egg-handling and poultry-cleaning rooms need general illumination. Supply one light fixture for every 100 to 150 square feet of floor area. These lights can be switched as a group. Work areas should have additional fixtures, separately controlled with wall switches. Provide two outlets for each ten feet of work space.

Mount convenience receptacles at 12 foot intervals around the room perimeter. Other receptacles will be needed after equipment positions are established.

Farm Shops and Machinery Sheds
Install general lighting outlets at 10 to 12 foot intervals and 4 to 6 feet from the wall. This will furnish one outlet for every 100 to 150 square feet of floor area. Wall switch the lights as a group. Three-way switching may be convenient if there is more than one doorway.

Work benches should have two lighting outlets for each 10 feet of length. Two 150-W reflector lamps (bulbs) mounted 50 inches above the bench give enough light. Fluorescent fixtures should be mounted in a continuous row at the same height over the front 1/2 of the bench.

Receptacles for general use circuits should be spotted every 4 feet along the work bench. At least one additional receptacle is needed on each wall, 44 inches from the floor.

Motors and other heavy amperage equipment need special attention. Specifics are in Chapter 6.

Machinery sheds need only one lighting outlet for every 200 to 300 square feet of floor space. Lights should be wall-switched. Exterior entrance lighting is a good idea in front of sheds. It should be installed and switched separately. Convenience receptacles can be spaced at 40-foot intervals along the walls. A machinery shed normally will not need any equipment connections. However, it may be handy to have a central outlet for a welder if large machinery repairs are planned.

Pumps
Chapter 6 explains how pump motors are wired. You'll also want to follow the manufacturer's recommendations. Any pump installed in a wet

area has to comply with the NEC rules on switching and wiring in a wet location.

Both the pump and its wiring should be remote from any fire hazard. The pump circuit needs a separate disconnect from the other farm wiring. That way, the pump can continue to operate during a fire even if other circuits are shut down. This special pump circuit should originate outside the building. The best place of origin is the yardpole with wiring running underground to the pump.

Both the pump and the motor have to be grounded. It should be protected with lightning arrestors either built-in or installed as recommended. When the pump is located in a pump house, include a light fixture and convenience outlet in your plan.

Standby Service

Standby generators are common on farms. There are two general types—engine-driven and tractor-driven. Engine-driven types are further classified as either automatic and non-automatic.

All types must be connected to the farm wiring system with a double-throw switch that disconnects the regular power service before the generator is connected. This is to prevent feeding power back into the main lines (and electrocuting linemen who are trying to restore power). It also prevents damage to the generator caused by power feedback when normal service is restored.

The amperage rating of the double-throw switch must be matched to the rating of the conductors supplying the normal load. The rating of the generator or of the load supplied when the generator is operating cannot be used. This switch doesn't have to be fused because the generator has its own set of fuses or breakers. The service is already protected through its own breaker panel.

Some areas require that the neutral conductor be switched as well as the live conductors. A three-pole, three-blade switch will be needed. Where this isn't necessary, the three-blade switch may have to be modified. Ask your power company for the exact switch requirements.

An automatic generator has a gasoline motor that starts automatically when the power fails. A double-throw magnetic switch normally feeds the wiring system from the power supplier's line. When the service is interrupted, the generator starts. The switch transfers the system to the generator automatically. The amp rating of the transfer control switch must be equal to that of the incoming service. Recommend to your customer that he test this automatic system regularly.

Non-automatic generator systems cost less and work just as well. The only difference is that they have to be started manually. The system is equipped with a "power off" alarm that alerts anyone nearby of the power failure. These alarms have both audible and visual signals. An indicator light must also be installed on the power line to show when power has been restored.

The system is switched back and forth manually between the main line and the generator with a double-throw switch. Again, start the engine regularly to make sure it will run when needed.

Tractor-driven units are the most economical sources of standby power. But it takes time and attention to hook up the drive belt and keep it aligned. This may make a difference in some situations. Wire a tractor-driven generator into a system the same way as a non-automatic generator.

The generator is usually connected to only critical pieces of equipment that will need emergency power. The smaller the load, the smaller and less expensive the generator. The owner will have to decide what equipment is essential during a power outage. It may be possible to stagger the use of this equipment to reduce the emergency load. Figures 5-15 through 5-17 show schematic diagrams of possible connections. Your power company may be able to supply more detailed drawings.

Be sure to match the electrical characteristics of the generator to the load. In most cases you'll install a 60-cycle AC unit which can operate 115/230 volt circuits. The capacity of the generator should be about three times the amp rating of the largest motor to be started. This margin is needed during motor starting. A smaller size generator can be used if it has a specially designed motor-starting capacity.

Another way to figure generator capacity is to allow 2 KW per brake HP of the largest motor. For example, a 6 KW generator should be used to start a 3 HP electric motor.

The horsepower supplied by the tractor or engine that drives the generator should be at least double the KW rating of the generator. For example, a 10 KW generator would need a 20 HP engine.

Generators that operate above 150 volts must be grounded. Balance the load on both coils of a single-phase generator. Three-phase generators need special wiring. The manufacturer should help you with the wiring for a system like this.

Yardpoles

Yardpoles are still used as a central distribution point on farm jobs. But any suitable building will

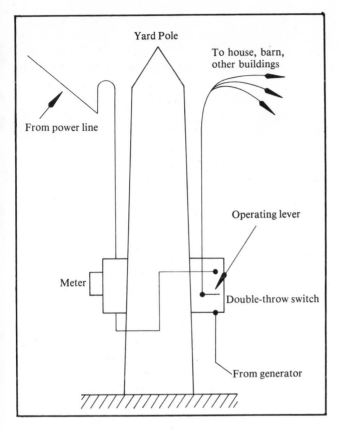

Stand-by Generator Connections—Self-contained Meter
Figure 5-15

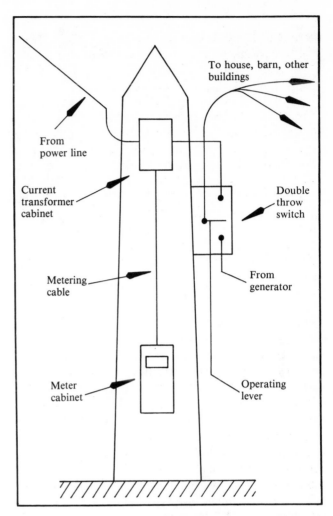

Stand-by Generator Connections—Current Transformer
Figure 5-16

serve nearly as well. Here's why a yardpole may be a good choice:

• Since the yardpole can be located near the electrical load center of the farm, feeder wire cost is kept to a minimum.

• Voltage drop is kept to a minimum.

• Wiring in one building can be done without interrupting service to other buildings.

• Service can be run directly from the pole to the pump house. This is an advantage in case of fire. Also, the loss of one building by fire is not likely to interrupt service to other buildings.

Locate the yardpole with the load of each building in mind. As a general rule, the heavier the load and the larger the distance between the pole and the building, the larger the wire required. Therefore, the yardpole should be closest to areas with the heaviest electrical loads.

The location of trees, driveways, and buildings may also influence location of the yardpole. You can't run distribution wires over or through buildings. Spend more than a few seconds considering the best location for the yardpole. Again, the power company may be able to help you here.

Sometimes you'll put switching equipment on the yardpole. Follow local practice and the suggestions of the utility company. Since each building served by the yardpole has its own service equipment, most utilities don't require a disconnect. Your customer is free to install one, however, if he wishes.

If used, this general disconnect switch should be weatherproof. It should also have a locking cover to prevent tampering even though a locked box presents a problem in an emergency.

Usually, the only other switches installed on a yardpole would be to permit standby service or a separate circuit for the pumphouse.

You'll probably mount a light on the top of the yardpole. This light can be switched manually on the pole or automatically with a photo cell.

A watthour meter is also mounted on the yardpole. Installation requirements vary with the utili-

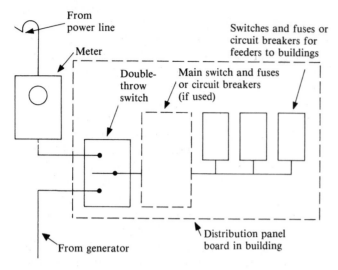

From
power line

Meter

Double-
throw
switch

Switches and fuses or
circuit breakers for
feeders to buildings

Main switch and fuses
or circuit breakers
(if used)

Distribution panel
board in building

From generator

Stand-by Generator Connections—Service Entrance
Figure 5-17

Residence	81 amps at 100% =	81 amps
Largest Demand (Poultry houses)	189 amps at 100% =	189 amps
Second Largest Demand (Feed Center)	36 amps at 75% =	27 amps
Third Largest Demand (Pump house)	12 amps at 65% =	8 amps
No additional demands	=	0
Computed demand at meter pole	=	305 amps

ty. One type of installation is the "self-contained" type (Figure 5-15). The full line current is brought down the pole to the meter socket and back up for distribution. This type of installation is usually restricted to 100 amp services but some utilities permit it for 200 amp services. Service cable can be used, but often conduit is required.

Conductor ratings for self-contained metering systems can be calculated with this formula:

1. 100% of Farm Residence Demand plus additional farm buildings, as follows:
2. 100% of largest computed demand
3. 75% of second largest computed demand
4. 65% of third largest computed demand
5. 50% of the sum of remaining demands of other loads

Chapter 6 explains how to compute demands. In the above formula, buildings or loads with the same function can be added together to find a single demand load size. For example:

A farm with a dwelling, four poultry houses, a feed center, and a pump house has the following computed loads.

Residence - 81 amps
Poultry House 1 - 101 amps
Poultry House 2 - 49 amps
Poultry House 3 - 25 amps
Poultry House 4 - 14 amps
Feed Center - 36 amps
Pump House - 12 amps

Adding all four poultry houses, the single demand load is 189 amps. Applying the formula, we have:

A high amperage system will need a current transformer installed at the pole top. (See Figure 5-16.) The line current passes through it and on to the various buildings. A metering current drawn in ratio to the line current is transferred from the transformer to the meter. Generally, the meter conductors are small. The exact size will be determined by the utility.

For large farms, you'll need more than one yard-pole for best economy and efficiency.

Conserving Electrical Energy

While conserving all types of energy is as important on the farm as it is in the city, farms offer more opportunities for conservation. Attention to efficiency is an important part of planning every wiring job. Don't skip this step. Use the best combinations of old and new energy technology that will do the job and still keep power costs to a minimum.

The following suggestions and recommendations can be incorporated into many electrical plans. Additional details and information are available from your local utility, state energy agencies, and agricultural schools and agencies.

Solar Power

There are two types of solar system—active and passive. A passive system uses power from the sun directly. It passes through windows, skylights, or open-front buildings. Where practical, a wall of stone or brick is used to absorb heat. Painting it black helps the absorption. When the temperature begins to drop, the wall releases the heat into the air.

An open-front livestock building facing south is a good example of a passive solar system. During the winter months, when the sun is low in the sky, the sunlight helps heat the building. In the summer, the sun is higher in the sky, so the building roof shades the front wall.

An active solar system collects and distributes heat from the sun by mechanical means such as blowers or pumps. Devices similar to the black wall in a passive solar system can be used in an active system to store the energy for later release at night or on cloudy days.

Active solar systems are gaining in popularity. Most farms can use large quantities of moderate-temperature air for grain drying, livestock ventilation and building heating. Farms also have large surface areas that are ideal for solar collectors. South walls and roofs of machinery sheds, barns, and single-story livestock buildings can all be used to mount the large collector panels used in an active system.

Solar collectors should be oriented to face the south for best efficiency. The precise alignment is determined by the latitude of the area. Even if the alignment isn't perfect, it may pay to install an active system.

Wind Power
Wind power is no stranger on the farm. Long before electricity was available in rural areas, wind was used for pumping water. Now, wind is used to furnish electricity. But don't overlook it for its original use where possible.

Wind generators are about 30% efficient in converting wind power to electrical power. Watts generated depend on propeller diameter, wind velocity, and the generator efficiency. For example, a two kilowatt generator with a 10 foot propeller diameter at an average wind speed of 12 miles per hour will generate 160 kilowatthours per month. An average household uses from 500 to 1,000 KWH per month.

Cost of wind generation is estimated to be between 10 and 15 cents per KWH, based on the initial investment and amortized over a 15-year period. That's more than power companies charge, but the gap is closing annually. But note that utilities are required by law to buy excess power from private generating sources. When the farm isn't using power and the wind generator is running, the power flows into the utility lines and is metered. When power is needed, it's drawn from the wind generator system until its capacity is exceeded. Then the power is drawn from the utility's lines. The customer ends up paying only the difference.

Electric Motors
Motors are wired to operate on single-phase or three-phase wiring. Three-phase motors are generally cheaper than comparable single-phase motors, but they need additional wiring. When deciding which type to recommend, consider both economy and intended use. Your electrical utility may be able to offer advice.

Whether the motor is single or three-phase, install the right size. An undersize or oversize motor costs more to operate than the correct rated size.

Rural power lines often have motor size limits based on motor starting currents. The starting current varies with the type of motor used. Soft-start motors are now available. These use less starting current than conventional motors.

More energy-efficient motors are also available. They use less energy, run at lower temperatures, and are less sensitive to line voltage changes. Overall, they provide cheaper service. Some power companies have penalty clauses which increase the rates for motors that operate at less than a given power factor. New motors generally operate at 80 to 85% power factor. Some of the older ones may operate at a power factor as low as 50%. Power Factor Correction Capacitors are available to correct this problem.

Proper maintenance increases motor efficiency. Motor and equipment mounts should be tight. Belt pulleys and V-belts should be replaced when worn. Proper belt tension is important. The motor should be kept dust and dirt free for cooler operation. Bearings should be in good condition and lubricated properly. Do not overlubricate. Make sure bearing and motor alignment is correct. Misalignment can cause overloading and a waste of energy. It is a good idea to occasionally clean switch contacts and brushes. Use fine sandpaper, not emery cloth.

Heat Exchangers and Heat Pumps
Since some farms use large amounts of artificial and natural heat, they are ideal for heat pumps and heat exchangers. Heat pumps powered by electricity are controlled by an automatic thermostat. In summer the unit uses circulating coils to extract heat from a building and release it outside. In winter, the opposite takes place.

Heating costs with a heat pump are generally more than heating with natural gas or oil but less than with electric heat. But heat pumps may be an economical choice wherever you need both heating and cooling.

Heat pumps work best in new construction where they can be used in a forced air system.

Heat exchangers use the hot air given off by animals or equipment to heat fresh air or water.

Heat exchangers work well in free stall barns, calf pens, and similar locations. Besides providing heat, they lower moisture and circulate fresh air. The energy a heat exchanger needs is used to drive a fan.

In dairy operations, heat given off in the milkhouse as a result of milk cooling can be used to heat hot water. At a cost of 5 cents per KWH, that saves over $30 a month in electric charges when 100 gallons of water is heated to 140 degrees F with a heat exchanger on a bulk tank compressor. Heat exchangers require a one-time investment. This investment may be paid back in three years.

Lighting

Always use the lowest possible bulb wattage that provides enough light for the job. Substitute one 100 watt incandescent lamp for two 60-watt lamps and you cut electric cost by 16% without changing the amount of light.

Although fluorescent fixtures cost more to install, they operate more cheaply and provide four times the light of incandescent lamps using the same watts. Outdoors, use mercury vapor lamps where possible. They provide twice as much light as an incandescent lamp per unit of energy used. If used in a cold climate, a low temperature unit is required.

Planned Loads

Planned load management is another way to reduce power costs. Electricity use can be controlled in commercial systems, industrial buildings and on the farm through careful planning. Smaller power units that operate over long periods of time are one way to lower wiring and equipment costs. For example, it may be cheaper to operate a 3 HP feed mill for ten hours a day than a 10 HP mill for three hours.

Some utilities charge a higher rate when electricity is used during peak periods. Work scheduling may make it possible to use heavy loads during low demand periods. Time-of-day metering is another method being used by some utilities. Under this plan, an appliance such as a water heater is wired so that a total of six hours of service is supplied daily during off-peak periods. The rate charged for this service is lower than the usual rate.

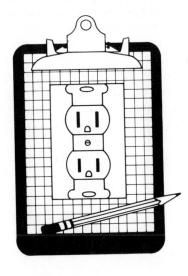

Chapter 6:

Electrical Design

Electrical design is done when planning is complete and before installation begins. Chapter 5 explained planning—deciding what fixtures, receptacles and switches will go where. At the design step, you lay out on paper the entire system so it will meet code and budget limitations.

In many respects, designing is like planning, just in more detail. The electrical design adds flesh and blood to the skeleton created by your plan. In some cases, you'll discover inconsistencies or potential problems at this point. Safety or economy may be better served by making changes in the plan. A good design resolves these conflicts before construction begins.

Designing is done with calculations. But no complex math is needed. We'll explain all you need to know for most installations in easy-to-understand steps. A calculator will save you time and help guard against mistakes.

Good forms and design tables make for good designs and fewer errors. Most electrical contractors use design forms and tables on every significant job. Have a set of forms ready for your own use and get in the habit of using them. Following a form reduces the chance of omitting some important step. This chapter has blank forms and tables that are reasonably complete. A printer in your area can reproduce any of these forms quickly and cheaply. But no form made up for some other contractor will fit your needs exactly. Change and improve these forms to fit the type of work you handle.

When the design is finished, details are transferred to the construction plans. Fixture and panel schedules are filled in. Outlets are circuited and the service equipment is labeled by size and type. This construction plan is the working drawing for installation, of course, but it is also the basis for cost estimates and will be submitted for issuance of the building permit.

Commercial Lighting

Planning and designing commercial lighting systems is really one step with two parts. It isn't practical to separate the two, so we'll cover both the planning and designing of commercial lighting systems together in the following section.

The primary consideration in commercial lighting is good illumination at an acceptable cost. For a job taken on time-and-materials, it's best to over-plan slightly—that is, furnish good general illumination and plenty of supplementary lighting as well. If the estimated job cost is too high, let the owner decide where to cut back.

For jobs where you submit a bid, it's more difficult. You have to balance good lighting design with low cost if you are doing the planning and design. Some customers will want to get by as cheaply as possible. Other customers will realize that good lighting increases production and building appearance. They will be willing to spend more for a better job. Check with the architect or the owner to find which direction he wants you to go.

Of course, if you are bidding on a professionally designed job, you won't be able to save costs through the design steps. You'll have to concentrate on furnishing materials and labor at low cost according to what's required in the plan and specs.

But a solid knowledge of what goes into the design work will help you prepare an accurate, low-cost bid.

Code Restrictions

The National Electric Code establishes minimum lighting standards and governs their installation. The standard is based on the minimum wattage allowed per square foot of building type. You also have to observe a maximum wattage. The minimum lighting loads by occupancy found in the NEC Section 220 are meant to guarantee enough lighting power. Maximum wattages have recently been included in some state and local codes. These values are intended to save energy by reducing wasteful lighting levels. Sometimes these values conflict, but local codes will have the final word. Check them carefully.

Footcandles

A *footcandle* is the illumination which falls on one square foot surface from a uniform point source called a *candela*. Table 6-1 shows recommended footcandles for illuminating different interior

Activity	Lighting Type	Average Footcandles (Per Square Foot)
Open room areas with no visual tasks performed; light traffic, dark areas acceptable.	General	3
Simple viewer orientation; no visual tasks performed: subdued lighting for passages, entertainment, social activity, closets, utility rooms and the like.	General	8
Areas where high contrast or large-sized visual tasks are performed occasionally; auditoriums, general lobbies, dining rooms, restrooms, large item storage, general loading areas.	General	15
Areas of steady high contrast, large-sized visual tasks: reading, writing areas, general work areas, simple assembly tasks and non-critical seeing areas, general shop areas.	Localized general	30
Areas of steady small-sized, medium-contrast work tasks: laboratories, commercial kitchens, mail sorting, sales transaction areas, moderately difficult assembly areas, drafting rooms and critical seeing areas.	Localized general/ Supplementary	75
Areas of steady, very small-sized, low-contrast tasks: hospital surgeries, difficult assembly areas, difficult inspection areas, color composing areas, fine instrument testing.	Localized general/ Supplementary	150
Areas of very small-sized, very low-contrast work tasks done over a long period: very difficult assembly or inspection, low contrast sewing, fabric inspection.	Supplementary with General background	300
Very small-sized, very low-contrast work tasks over a very long period of time.	Supplementary with General background	750

Recommended Interior Footcandles
Table 6-1

areas. These are average values. Although handy to use as a guideline, footcandles cannot be the only consideration when planning a lighting system. Other important variables must be brought into your planning. The speed and accuracy of the task, the age of the workers, the reflectance of the physical surroundings, the efficiency of the light fixtures and the ratio of general illumination to task lighting should all be included in your final plan. Each of these will be dealt with in this chapter.

Lumens, Lamps and Luminaires

Lamp is an electrician's term for what most people call a "bulb." The efficiency of a lamp is measured in *lumens*. For example, a candle gives off about 0.5 lumens per watt. A 60-watt tungsten filament lamp—the common household bulb—gives off about 14 lumens per watt. A 40-watt fluorescent lamp is more efficient. It gives off about 73 lumens per watt.

A *luminaire* is an electrician's term for a lighting fixture. It includes the lamp, the reflector (if any), the mounting device, and all electrical connections within the unit. Since the reflector or parts of the luminaire that distribute light can affect the efficiency of the lamp, luminaires are given a lumen rating of their own by the manufacturer.

It's important to understand each of these terms to avoid confusion when calculating lighting systems.

Types of Lighting

There are three broad types of lighting: *general, localized general* and *supplementary*.

General lighting is the kind found in buildings where an even amount of illumination is spread over a wide area. Examples of general lighting would be in lobbies of public buildings, classrooms, and large offices.

Localized general lighting serves two purposes. The light source is intended to illuminate a work or task area but also furnishes general lighting for the surrounding area. A light fixture over a kitchen table would be a common example of this type of lighting.

Supplementary lighting illuminates a particular task or need. Any time supplementary lighting is used, you must provide general or localized general lighting for eye comfort. The ratio of brightness of supplementary light to general light should not be greater than 10 to 1. Supplementary lighting is used in work and task areas where people read, or work with small items.

Types of Luminaires

Luminaires are divided into five types (Figure 6-2): direct, semi-direct, general diffuse, semi-indirect, and indirect.

These same names are sometimes used to describe a lighting system made up of one of these types of luminaires.

Direct - A luminaire designed to distribute 90% or more of its light downward below the horizontal mounting plane. Direct luminaires can be suspended, mounted on the ceiling or recessed flush (Figure 6-3) with the ceiling. They are either concentrating or distributing type.

Concentrated luminaires direct much of their light to a work area without much loss to the walls, windows and ceiling. Lighting systems of this type illuminate horizontal planes such as table tops much more strongly than vertical planes such as walls.

A distributing luminaire is designed to spread the direct light over a wider area. They provide enough light when mounted as high as 10 to 12 feet above floor level. They must not be spaced more than the mounting height above the floor.

Direct systems are usually found in shops and factories where work and work surfaces are not bright and reflective. They are not recommended for applications such as offices or classrooms because they produce glare.

Semi-direct - A luminaire that distributes between 60 and 90% of its light downward below the horizontal plane and 10 to 40% upward. It may be suspended or ceiling mounted. The light source is either exposed, concealed behind a glass or plastic or behind a louver. These units give mostly direct light. Some indirect light is available if the ceiling is a light color. Any indirect light adds to eye comfort by reducing sharp contrast between the light source and its background. It also improves ceiling appearance.

Semi-direct units are similar to direct units but are used primarily where shadows and glare will not be a problem. They can be used in offices and classrooms if carefully positioned to direct reflected images away from the eye. When used in offices, the room, equipment and task surfaces should have a low gloss finish.

General Diffuse - This luminaire distributes 40 to 60% of its light upward from the horizontal. It can be suspended or mounted on a wall, ceiling or pedestal. The lamp is enclosed in a diffusing enclosure of glass or plastic.

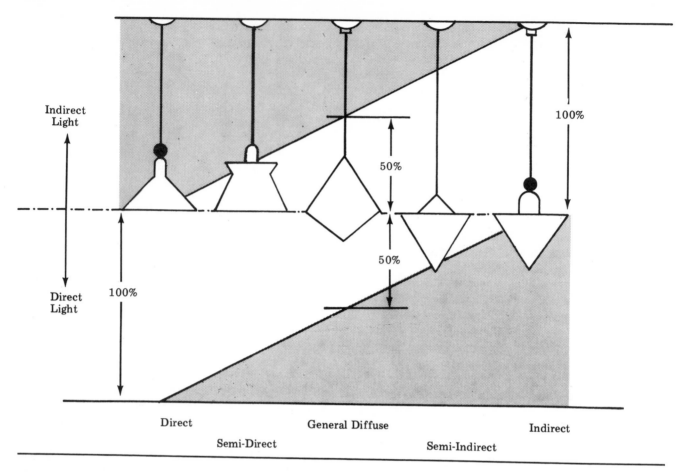

Indirect
Light

Direct
Light

100%

50%

50%

100%

Direct

Semi-Direct

General Diffuse

Semi-Indirect

Indirect

Basic Luminaires
Figure 6-2

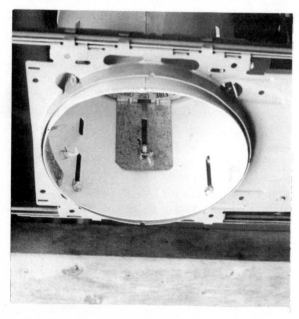

A Roughed-In Recessed Direct Luminaire
Figure 6-3

The most common example is the simple glass globe (Figure 6-4) with a filament lamp. It's a simple, low-cost fixture that gives good illumination on walls and other vertical surfaces. Usually the manufacturer sets a maximum wattage for the lamp that can be enclosed in the globe because of heat build-up. If not, you can get more light by using a larger lamp. If this produces too much glare, the globe size may have to be increased.

Semi-indirect - A luminaire that distributes 60 to 90% of its light above the horizontal and 10 to 40% below. The light source is hidden from view below the horizontal by diffusing glass or plastic. Because reflected ceiling light makes up an important part of the system, the ceiling must be light colored but not overly bright. It should be kept clean and painted.

Indirect - Indirect luminaires distribute 90 to 100% of their light upward above the horizontal. They are suspended or mounted on a wall or pedestal.

Courtesy: Thomas Industries

A Decorative Globe Fixture
Figure 6-4

The light source is hidden from view below the horizontal. The ceiling is really the light source. Shadows and reflected glare are small. If the light intensity isn't too high, there's very little glare.

Indirect systems aren't used if more than 75 footcandles of horizontal illumination are needed. Higher footcandles cause uncomfortable ceiling brightness, and direct or reflected glare.

Any of these five luminaires can use either fluorescent or incandescent lamps. However, at levels above 35 footcandles, good illumination quality becomes difficult with incandescent lamps.

Also, incandescent lamps give off a higher percentage of heat than fluorescent lamps. This may be an advantage or a disadvantage, depending on the climate. Initial installation cost of incandescents is less than fluorescent. But fluorescents are cheaper to operate over the long run.

The final choice of which lamp and luminaire to use should be based on four considerations:
1. The work or activity taking place in the area
2. The composition of the area
3. The overall appearance desired
4. Cost

Calculating a Lighting System
You can design a simple lighting system with the information in a fixture manufacturer's catalog.

Wattage and footcandles for their luminaires will be in the catalog. A light distribution curve will also be shown. This indicates how the light falls. Pictures and sketches show the appearance of the luminaire and its mounting. A spacing to mounting height ratio will also be given so you know how to place the luminaire. The catalog may also indicate the list price.

In a general use area such as a warehouse or building lobby, there's enough information in the catalog to design a satisfactory system. You may be able to call on the manufacturer's salesmen for help. Often they have special training in lighting design.

But when lighting is critical, you should know how to do the calculations and verify that the system you install will be what the owner needs. The next few pages explain the most common method of calculating fixture lighting.

The Zonal Cavity Method
This method of light calculation divides the room into three zones: the *ceiling cavity* extends from the light fixture level to the ceiling; the *room cavity* is the space between light fixture level and the work level; the *floor cavity* lies between the floor and the work level.

Differences in reflective value of the three areas is determined and given a numerical value. Other values are calculated for items such as light maintenance and the luminaire design. These numbers are then used in a final formula to find the number of light fixtures needed.

All of this is much easier than it sounds when you use a worksheet like Figure 6-5. If more than one room is calculated, use a summary sheet like Figure 6-6.

Step 1 - Room Dimensions Record the inside room dimensions either from the plans or by measuring the room. Round them off to the nearest inch. Multiply the length times the width to find the floor area. Note the ceiling height and proposed fixture height.

Step 2 - Reflectance Factors Reflectance indicates the amount of light reflected by room surfaces. Generally, a hard white surface will reflect the most light. Darker and textured surfaces reflect less. Ceiling height, exposed joists and ductwork also affect reflectance. Ceilings have a greater reflectance value than walls. Ceilings' values are 90, 80, 70, 50, 30 and 10%, with 90% the maximum. Walls have reflectances of 70, 50 and 30%. Floors are generally rated at 20% or 10%. Reflec-

No. _____

Project _____ Date _____

Room _____ Footcandles/Sq. Ft. _____

Room Dimensions

Length _____
Width _____
Floor area _____
Ceiling height _____
Fixture height _____

Reflectance

Wall _____ %
Ceiling _____ %
Floor _____ %

Fixture Data

Mfr. _____
Cat. No. _____
Lamps/Fix. _____
Lumens/Lamp _____
Coeff. of Utilization (CU) _____
Maintenance Factor (MF) _____

Cavity Data

Room cavity
  Height _____
  Ratio _____

Ceiling cavity
  Height _____
  Ratio _____
  Eff. Reflectance _____

Floor cavity
  Height _____
  Ratio _____
  Eff. Reflectance _____

Ceiling

h cc

h rc

Fixture height

h fc

Workplane

h

Floor

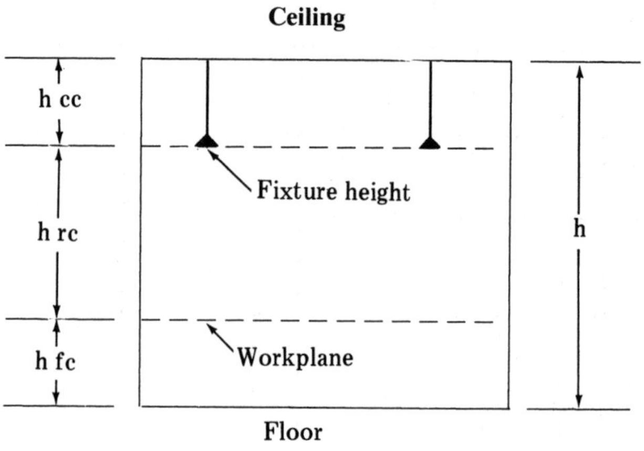

$$\frac{5 \times \text{cavity height} \times (\text{length} + \text{width})}{\text{Length} \times \text{width}} = \text{Cavity Ratio}$$

$$\frac{5 \times \quad \times (\quad + \quad)}{\quad \times} = \quad = \text{RCR} \qquad \frac{5 \times \quad \times (\quad + \quad)}{\quad \times} = \quad = \text{CCR}$$

$$\frac{5 \times \quad \times (\quad + \quad)}{\quad \times} = \quad = \text{FCR}$$

$$\frac{\text{Floor area} \times \text{footcandles}}{\text{Lamps/Fix.} \times \text{Lumens/Lamp} \times \text{CU} \times \text{MF}} = \text{No. of Fixtures}$$

$$\frac{\quad \times \quad}{\quad \times \quad \times \quad} =$$

$$\frac{\text{No. of Fixtures} \times \text{Lamps/Fix.} \times \text{Lumens/Lamp} \times \text{CU} \times \text{MF}}{\text{Floor area}} = \text{Footcandles}$$

$$\frac{\quad \times \quad \times \quad \times \quad}{\quad} =$$

Illumination Calculation Sheet
Figure 6-5

Building Area:

Room No.	Reflec. Clg./ Wall/Flr.	Ceiling Type	Length	Width	Height	HRC	RCR	Fixture Type	Lumens/ Fixt.	CU	LLF	FC/ Fixt.	Design FC	No. Fix. Install	FC	Total Ltg. Watts	Ltg. Watts/ Sq. Ft.

Lighting Summary Sheet
Figure 6-6

tance values can measured exactly with a reflec-tometer. Another method is to match color samples of known reflectance with the surface to be lighted.

Step 3 - Fixture Data Records information about the luminaire which you intend to use. This infor-mation is given in the manufacturer's catalog. It will include the Maintenance Factor (MF). This number is sometimes called the Light Loss Factor (LLF). It shows the predicted decline in light caus-ed by the aging of lamps and dirt settling on the fix-tures and reflective surfaces. It's also affected by the rate of lamp replacement and changes of input voltage. Three categories can be used to determine Maintenance Factors:

1. *Good* - Air is generally clean, luminaires are cleaned regularly. Lamps are replaced periodically as a group.

2. *Medium* - Air quality is fair, luminaires are cleaned occasionally. Lamps are replaced only when they burn out.

3. *Poor* - Air is very dirty. Lamps and luminaires receive attention only irregularly.

The type of use, an inspection of the building or a talk with the building owner or manager will in-dicate which category applies. After estimating the correct category, find the maintenance factor in the manufacturer's catalog. The Illuminating Engineering Society (IES) Handbook also gives tables and charts for determining maintenance fac-tors for various types of fixtures.

Another number you need is the *coefficient of utilization* (CU). The CU takes into account the fixture design and how it will interact with the room surfaces. The manufacturer will have a table which gives the CU for a wide variety of reflec-tances.

The IES Handbook also lists coefficient of utilization tables for many fixtures. CU values de-pend on *effective reflectance*. Effective reflectance is found by taking the ceiling or floor cavity ratio and applying it along with the surface reflectance to an effective reflectance table. This takes place in Step 5.

Step 4 - Cavity Data Cavity data considers the height of the room, ceiling and floor cavities. Measure all height data from the floor, not the ceil-ing. Room cavity height can be found by subtrac-ting the work plane height from the fixture moun-ting height. The normal *work plane* is 30 inches (2.5 ft.) from the floor. Some areas may use other heights. The ceiling cavity height is the *room* height (not the room cavity height), less the fixture moun-

ting height. If fixtures are ceiling mounted, there will be no ceiling cavity. The floor cavity height is always the same as the work plane.

You use the cavity ratios and the effective reflec-tance for the ceiling and floor cavities in Step 5. The room cavity, since it does not have horizontal surfaces, has little effective reflectance.

Step 5 - Cavity Ratios (CR) Cavity ratios are calculated with the formula:

$$\text{Cavity Ratio} = \frac{5 \times \text{cavity height} \times (\text{length} + \text{width})}{\text{length} \times \text{width}}$$

When the ratios have been calculated, find the ef-fective reflectance of ceiling and floor cavities with Table 6-7. The ratios and effective reflectance values are now entered into the Cavity Data sec-tion. These values are then used to find the Coeffi-cient of Utilization from the manufacturer's tables. Record this figure in the Fixture Data section of the work sheet. Note that the Room Cavity Ratio is not used to find effective reflectance. It is used, however, in the Coefficient of Utilization table.

Step 6 - Calculating the Number of Fixtures Now that the necessary figures are on hand, apply the formula:

$$\text{No. Fixtures} = \frac{\text{floor area} \times \text{desired footcandles}}{\text{lamps per fix.} \times \text{lumens per lamps} \times \text{CU} \times \text{MF}}$$

If you need to find the light level that will be pro-duced by a certain number of fixtures, the formula is rearranged:

$$\text{Footcandles} = \frac{\begin{array}{c}\text{No. of fix.} \times \text{lamps per fix.} \times \\ \text{lumens per lamps} \times \text{CU} \times \text{MF}\end{array}}{\text{floor area}}$$

Step 7 - Fixture Location Where you mount fix-tures depends on the shape of the room and where work will be done. But don't exceed the spacing-to-mounting height ratios given by the manufacturer. Keep in mind the following:

These ratios are maximums. It may be necessary to mount luminaires closer together to get enough light where you need it. Since lamps are cheaper than luminaires, you may be tempted to cut costs by mounting high-wattage lamps at the maximum

Effective Ceiling or Floor Reflectance
Table 6-7

Per Cent Base* Reflectance → Cavity Ratio ↓ (Per Cent Wall Reflectance)	50										60										70										80										90									
Wall →	90	80	70	60	50	40	30	20	10	0	90	80	70	60	50	40	30	20	10	0	90	80	70	60	50	40	30	20	10	0	90	80	70	60	50	40	30	20	10	0	90	80	70	60	50	40	30	20	10	0
0.2	50	50	49	49	48	48	47	46	46	44	60	59	59	58	57	56	56	55	55	53	70	69	68	68	67	66	66	66	65	64	79	78	78	77	77	76	76	75	74	72	89	88	88	87	86	85	85	84	84	82
0.4	50	49	48	48	47	46	45	45	44	42	60	59	58	57	55	54	53	52	52	50	69	68	67	66	65	64	63	62	61	58	79	77	76	75	74	73	72	71	70	68	88	87	86	84	83	81	80	79	79	76
0.6	50	48	47	46	45	44	43	42	41	38	60	58	57	56	55	53	51	51	50	46	69	67	65	64	63	61	59	58	57	54	78	76	75	73	71	70	68	66	65	63	87	86	84	82	80	79	77	76	74	73
0.8	50	48	47	45	44	42	40	39	38	36	59	57	56	54	53	51	48	47	46	43	68	66	64	62	60	58	56	55	53	50	78	75	73	71	69	67	65	63	61	57	87	85	82	80	77	75	73	71	69	67
1.0	50	48	46	44	43	41	38	37	36	34	59	57	55	53	51	48	45	44	43	41	68	65	62	60	58	55	53	52	50	47	77	74	72	69	67	65	62	60	57	55	86	83	80	77	75	72	69	66	64	62
1.2	50	47	45	43	41	39	36	35	34	29	59	56	54	51	49	46	44	42	40	38	67	64	61	59	57	54	50	48	46	44	76	73	70	67	64	61	58	55	53	51	85	82	78	75	72	69	66	63	60	57
1.4	50	47	45	42	40	38	35	34	32	27	59	56	53	49	47	44	41	39	38	36	67	63	60	58	55	51	47	45	44	41	76	72	68	65	62	59	55	53	50	48	85	80	77	73	69	65	62	59	57	52
1.6	50	47	44	41	39	36	33	32	30	26	59	55	52	48	45	42	39	37	35	33	67	62	59	56	53	49	45	43	41	38	75	71	67	63	60	57	53	50	47	44	84	79	75	71	67	63	59	56	53	50
1.8	50	46	43	40	38	35	31	30	28	25	58	55	51	47	44	40	37	35	33	31	66	61	58	54	51	46	42	40	38	35	75	70	66	62	58	54	50	47	44	41	83	78	73	69	64	60	56	53	50	48
2.0	50	46	43	40	37	34	30	28	26	24	58	54	50	46	43	39	35	33	31	29	66	60	56	52	49	45	40	38	36	33	74	69	64	60	56	52	48	45	41	38	83	77	72	67	62	56	53	50	47	43
2.2	50	46	42	38	36	33	29	27	24	22	58	53	49	45	42	37	34	31	29	28	66	60	55	51	48	43	38	36	34	32	74	68	63	58	54	49	45	42	38	35	82	76	70	65	59	54	50	47	44	40
2.4	50	46	42	37	35	31	27	25	23	21	58	53	48	44	41	36	32	30	27	26	65	60	54	50	46	41	37	35	32	30	73	67	61	56	52	47	43	40	36	33	82	75	69	64	58	53	48	45	42	37
2.6	50	46	41	37	34	30	26	23	21	20	58	53	48	43	39	35	31	28	26	24	65	59	54	49	45	40	35	33	30	28	73	66	60	55	50	45	41	38	34	31	81	74	67	62	56	51	46	42	38	35
2.8	50	46	41	36	33	29	25	22	20	19	58	53	47	43	38	34	29	27	24	22	65	59	53	48	43	38	33	30	28	26	73	65	59	53	48	43	39	36	32	29	81	73	66	60	54	49	44	40	36	34
3.0	50	45	40	36	32	28	24	21	19	17	57	52	46	42	37	32	28	25	23	20	64	58	52	47	42	37	32	29	27	24	72	65	58	52	47	42	37	34	30	27	80	72	64	58	52	47	42	38	34	30
3.2	50	44	39	35	31	27	23	20	18	16	57	51	45	41	36	31	27	23	22	18	64	58	51	46	40	36	31	28	25	23	72	65	57	51	45	40	35	33	28	25	79	71	63	56	50	45	40	36	32	28
3.4	50	44	39	35	30	26	22	19	17	15	57	51	45	40	35	30	26	23	20	17	64	57	50	45	39	35	29	27	24	22	71	64	56	49	44	39	34	32	27	24	79	70	62	54	48	43	38	34	30	27
3.6	50	44	39	34	29	25	21	18	16	14	57	50	44	39	34	29	25	22	19	16	63	56	49	44	38	33	28	25	22	20	71	63	54	48	43	38	32	30	25	23	78	69	61	53	47	42	36	32	28	25
3.8	50	44	38	34	29	25	21	17	15	13	57	50	43	38	33	29	24	21	19	15	63	56	49	43	37	32	27	23	21	19	70	62	53	47	41	36	31	28	24	22	78	69	60	51	45	40	35	31	27	23
4.0	50	44	38	33	28	24	20	17	15	12	57	49	42	37	32	28	23	20	18	14	63	55	48	42	36	31	26	23	20	17	70	61	53	46	40	35	30	26	22	20	77	68	58	51	44	39	33	29	25	22
4.2	50	43	37	32	28	24	20	17	14	12	56	49	42	37	32	27	22	19	17	14	62	55	47	41	35	30	25	22	19	16	69	60	52	45	39	34	29	25	21	18	77	67	57	50	43	37	32	28	24	21
4.4	50	43	37	32	27	23	19	16	13	11	56	49	42	36	31	27	22	19	16	13	62	54	46	40	34	29	24	21	18	15	69	60	51	44	38	33	28	24	20	17	76	66	56	49	42	36	31	27	23	20
4.6	50	43	36	31	26	22	18	15	13	10	56	49	41	35	30	26	21	18	16	13	62	54	46	39	33	28	24	21	17	14	69	59	50	43	37	32	27	23	19	15	76	65	55	47	40	35	30	26	22	19
4.8	50	43	36	31	26	22	18	15	12	09	56	48	41	34	29	25	21	18	15	12	62	53	45	38	32	27	23	20	16	13	68	58	49	42	36	31	26	22	18	14	75	64	54	46	39	34	29	25	21	18
5.0	50	42	35	30	25	21	17	14	12	09	56	48	40	34	28	24	20	17	14	11	61	52	44	36	31	26	22	19	16	12	68	58	48	41	35	30	25	21	18	14	75	63	53	45	38	33	28	24	20	16
6.0	50	42	34	29	23	19	15	13	10	06	55	45	37	31	25	21	17	14	11	07	60	51	41	35	28	24	19	16	13	09	66	55	44	38	31	27	22	19	15	10	73	61	49	41	34	29	24	20	16	11
7.0	49	41	32	27	21	18	14	11	08	05	54	43	35	30	24	20	15	12	09	05	58	48	38	32	26	22	17	14	11	06	64	53	41	35	28	24	19	16	12	07	70	58	45	38	30	27	21	18	14	08
8.0	49	40	30	25	19	16	12	10	07	03	53	42	33	28	22	18	14	11	08	04	57	46	35	29	23	19	15	13	10	05	62	50	38	32	25	21	17	14	11	05	68	55	42	35	27	23	18	15	12	06
9.0	48	39	29	24	18	15	11	09	07	03	52	40	31	26	20	16	12	10	07	03	56	45	33	27	21	18	14	12	09	04	61	49	36	30	23	19	15	13	10	04	66	52	38	31	25	21	16	14	11	05
10.0	47	37	27	22	17	14	10	08	06	02	51	39	29	24	18	15	11	09	07	02	55	43	31	25	19	16	12	10	08	03	59	46	33	27	21	18	14	11	08	03	65	51	36	29	22	19	15	11	09	04

* Ceiling, floor, or floor of cavity.

Effective Ceiling or Floor Reflectance
Table 6-7 (continued)

* Ceiling, floor, or floor of cavity.
From the IES Lighting Handbook, 1981 Reference Volume

Per Cent Base Reflectance = 40

Per Cent Wall Reflectance →	90	80	70	60	50	40	30	20	10	0
Cavity Ratio 0.2	40	40	39	39	38	38	37	37	36	36
0.4	41	40	39	38	37	36	35	35	34	34
0.6	41	40	39	38	37	36	34	33	32	31
0.8	41	40	38	37	36	35	33	32	31	29
1.0	42	40	38	37	35	33	32	31	29	27
1.2	42	40	38	36	34	32	30	29	27	25
1.4	42	39	37	35	33	31	29	27	25	23
1.6	42	39	37	35	32	30	27	26	25	22
1.8	42	39	36	34	31	28	26	24	24	21
2.0	42	39	36	34	31	28	25	23	23	19
2.2	42	39	36	33	30	27	24	22	22	18
2.4	43	39	35	33	29	27	24	21	21	17
2.6	43	39	35	32	29	26	23	20	20	15
2.8	43	39	35	32	28	25	22	19	19	14
3.0	43	39	35	31	27	24	21	18	18	13
3.2	43	39	35	31	27	23	20	17	17	13
3.4	43	39	34	30	26	23	20	17	17	12
3.6	44	39	34	30	26	22	19	16	16	11
3.8	44	38	33	29	25	22	19	16	16	10
4.0	44	38	33	29	25	21	18	15	15	10
4.2	44	38	33	29	24	21	17	15	15	10
4.4	44	38	33	28	24	20	17	14	14	09
4.6	44	38	32	28	23	19	16	14	14	08
4.8	44	38	32	27	22	19	16	13	13	07
5.0	45	38	31	27	22	19	15	13	13	07
6.0	44	37	30	25	20	17	13	11	08	05
7.0	44	36	29	24	19	16	12	10	07	04
8.0	44	35	28	23	18	15	11	09	06	03
9.0	44	35	26	21	16	13	09	07	05	02
10.0	43	34	25	20	15	12	08	07	05	02

Per Cent Base Reflectance = 30

Per Cent Wall Reflectance →	90	80	70	60	50	40	30	20	10	0
Cavity Ratio 0.2	31	30	30	29	29	29	28	28	28	27
0.4	31	31	30	30	29	28	28	27	26	25
0.6	32	31	30	29	28	27	26	25	25	23
0.8	32	31	30	29	28	26	25	23	23	22
1.0	33	32	30	29	27	25	24	23	22	20
1.2	33	32	30	28	27	25	23	22	21	19
1.4	34	32	30	28	26	24	22	21	20	18
1.6	34	33	29	27	25	23	22	20	18	17
1.8	35	33	29	27	25	23	21	19	17	16
2.0	35	33	29	26	24	22	20	18	16	14
2.2	36	32	29	26	24	22	19	17	15	13
2.4	36	32	29	26	24	22	19	16	14	12
2.6	36	32	29	25	23	21	18	16	14	12
2.8	37	33	29	25	23	21	17	15	13	11
3.0	37	33	29	25	22	20	17	15	12	10
3.2	37	33	29	25	22	19	16	14	12	10
3.4	37	33	29	25	22	19	16	14	11	09
3.6	38	33	29	24	21	18	15	13	10	09
3.8	38	33	28	24	21	18	15	13	10	08
4.0	38	33	28	24	21	18	14	12	09	07
4.2	38	33	28	24	20	17	14	12	09	07
4.4	39	33	28	24	20	17	14	11	09	06
4.6	39	33	28	24	20	17	13	10	08	05
4.8	39	33	28	24	20	16	13	10	08	05
5.0	39	33	28	24	19	16	13	10	08	05
6.0	39	33	27	23	18	15	11	09	06	04
7.0	40	33	26	22	17	14	10	08	05	03
8.0	40	33	26	21	16	13	09	07	04	02
9.0	40	33	25	20	15	12	09	07	04	02
10.0	40	32	24	19	14	11	08	06	03	01

Per Cent Base Reflectance = 20

Per Cent Wall Reflectance →	90	80	70	60	50	40	30	20	10	0
Cavity Ratio 0.2	21	20	20	20	20	19	19	19	19	17
0.4	22	21	20	20	19	19	19	18	17	16
0.6	23	21	21	20	20	19	18	17	16	15
0.8	24	22	21	20	19	18	17	16	15	14
1.0	25	23	22	21	20	18	17	16	15	13
1.2	25	23	22	20	19	17	16	15	13	12
1.4	26	24	22	20	18	17	16	15	14	12
1.6	26	24	22	20	18	17	16	15	13	11
1.8	27	25	23	21	19	17	15	14	13	10
2.0	28	25	23	20	18	16	15	13	11	09
2.2	28	25	23	20	18	16	14	12	11	09
2.4	29	26	23	20	18	16	14	12	10	08
2.6	29	26	23	20	18	16	14	12	10	08
2.8	30	27	23	20	18	15	13	11	09	07
3.0	30	27	23	20	17	15	13	11	09	07
3.2	31	27	23	20	17	15	12	11	09	06
3.4	31	27	23	20	17	15	12	10	09	06
3.6	32	27	23	20	17	15	12	10	08	05
3.8	32	28	23	20	17	14	12	10	08	05
4.0	33	28	23	20	17	14	11	09	07	05
4.2	33	28	24	20	17	14	11	09	07	04
4.4	34	28	24	20	17	14	11	09	07	04
4.6	34	29	24	20	16	13	11	09	07	04
4.8	34	29	24	20	16	13	10	08	06	04
5.0	35	29	24	20	16	13	10	08	06	04
6.0	36	30	24	20	16	13	10	08	05	02
7.0	36	30	24	20	15	12	09	07	04	02
8.0	37	30	23	19	15	12	08	06	03	01
9.0	37	29	23	19	14	11	08	06	03	01
10.0	37	29	22	18	13	10	07	05	03	01

Per Cent Base Reflectance = 10

Per Cent Wall Reflectance →	90	80	70	60	50	40	30	20	10	0
Cavity Ratio 0.2	11	11	11	10	10	10	09	09	09	09
0.4	12	11	11	11	10	10	09	09	09	08
0.6	13	13	12	11	11	10	09	09	08	08
0.8	15	14	13	12	11	10	09	09	08	07
1.0	16	14	13	12	11	10	09	08	08	07
1.2	17	15	14	13	12	11	10	09	07	06
1.4	18	16	14	13	12	11	10	09	07	06
1.6	19	17	15	14	13	11	10	09	07	06
1.8	19	17	15	14	13	11	09	07	06	05
2.0	20	18	16	14	13	11	09	08	06	05
2.2	21	19	16	14	13	11	09	07	06	05
2.4	22	19	17	15	13	11	09	07	06	05
2.6	23	20	17	15	13	11	09	07	06	04
2.8	23	20	18	16	13	11	09	07	05	03
3.0	24	21	18	16	13	11	09	07	05	03
3.2	25	21	18	16	13	11	09	07	05	03
3.4	26	22	19	16	14	11	09	07	05	03
3.6	26	22	19	16	14	11	09	06	04	03
3.8	27	23	19	17	14	11	09	06	04	02
4.0	27	23	20	17	14	11	09	06	04	02
4.2	28	24	20	17	14	11	09	06	04	02
4.4	28	24	20	17	14	11	08	06	04	02
4.6	29	25	20	17	14	11	08	06	04	02
4.8	29	25	20	17	14	11	08	06	04	02
5.0	30	25	20	17	14	11	08	06	04	02
6.0	31	26	21	18	14	11	08	06	03	01
7.0	32	27	21	17	13	11	08	06	03	01
8.0	33	27	21	17	13	10	07	05	03	01
9.0	34	28	21	17	13	10	07	05	02	01
10.0	34	28	21	17	12	10	07	05	02	01

Per Cent Base Reflectance = 0

Per Cent Wall Reflectance →	90	80	70	60	50	40	30	20	10	0
Cavity Ratio 0.2	02	02	02	01	01	01	01	00	00	0
0.4	04	03	03	02	02	02	01	01	00	0
0.6	05	05	04	03	03	02	02	01	01	0
0.8	07	06	05	04	04	03	02	02	01	0
1.0	08	07	06	05	04	03	02	02	01	0
1.2	10	08	07	06	05	04	03	02	01	0
1.4	11	09	08	07	06	04	03	02	01	0
1.6	12	10	09	07	06	05	03	02	01	0
1.8	13	11	09	08	07	05	04	03	01	0
2.0	14	12	10	09	07	05	04	03	01	0
2.2	15	13	11	09	07	06	04	03	01	0
2.4	16	13	11	09	08	06	04	03	02	0
2.6	17	14	12	10	08	06	05	03	02	0
2.8	17	15	13	10	08	07	05	03	02	0
3.0	18	16	13	11	09	07	05	03	02	0
3.2	19	16	14	11	09	07	05	03	02	0
3.4	20	17	14	12	09	07	05	03	02	0
3.6	20	17	15	12	10	08	05	04	02	0
3.8	21	18	15	12	10	08	05	04	02	0
4.0	22	18	15	13	10	08	05	04	02	0
4.2	22	19	16	13	10	08	06	04	02	0
4.4	23	19	16	13	10	08	06	04	02	0
4.6	23	20	17	13	11	08	06	04	02	0
4.8	24	20	17	14	11	08	06	04	02	0
5.0	25	21	17	14	11	08	06	04	02	0
6.0	27	23	18	15	12	09	06	04	02	0
7.0	28	24	19	15	12	09	06	04	02	0
8.0	30	25	20	15	12	09	06	04	02	0
9.0	31	25	20	15	12	09	06	04	02	0
10.0	31	25	20	15	12	09	06	04	02	0

spacing. This causes excessive glare, especially if the ceiling and wall are highly reflective.

Generally, put fluorescent luminaires in parallel rows in large areas. If there is a definite line of sight, the luminaires should run parallel to it. If work areas are located next to the walls, outer rows should take this into account. In small areas, fluorescents should be mounted in "U" or "L" patterns. This usually puts lighting over specific work areas. These patterns also provide better diffusion than parallel rows.

Keep indirect and semi-indirect luminaires far enough from the ceiling to keep ceiling glare to a minimum. Two-lamp fluorescents need at least 12 inches; four-lamp units, 24 inches. Indirect incandescent luminaires with 500-watt lamps should be suspended at least 20 inches below the ceiling; 750-watt lamps need 24 inches

The last step after locating the fixtures is to draw them on the electrical plan. If there isn't enough room on the floor plan to show all fixtures, use a separate sheet. Finally, transfer fixture information from the fixture data column of the illumination calculation sheet to the fixture schedule of the electrical plan.

Interpolation

Tables used for these zonal cavity calculations don't cover every possible value. Most of the time the number you need will fall between two numbers listed in the table. When this happens, you have to find an appropriate value between the two values in the table. This is called interpolation.

For example, in Effective Reflectance Table 6-7, no values are listed for a 15% Base Reflectance. Assume you have a room whose calculated floor reflectance is 15%. Its walls have a 50% reflectance and the floor cavity ratio is 1:2.

Look at the column in Table 6-7 that shows a 20% floor reflectance, a 50% wall reflectance and a cavity ratio of 1:2. The value is 19%. For a room with 10% floor reflectance and identical wall reflectance and cavity ratio, the value is 12%. Since 15% floor reflectance lies midway between 10 and 20%, we can assume that the effective reflectance value lies midway between 12 and 19%.

19 - 12   =   7

7/2   =   3.5 = 4 (rounded off)

12 + 4   =   16%

Sixteen percent, therefore, is the correct effective reflectance for a cavity with a 15% floor reflectance and a 50% wall reflectance. Use interpola-

tion on any lighting table wherever necessary. However, if the calculated value is within one or two numbers of a number on a table, use that number rather than interpolating. Small differences won't make a noticeable difference. Likewise, if a value is near midway between two numbers on the table, use the midway point for your calculations. That makes interpolation easier.

Example 1 - A General Office An electronics inspection lab where the work tasks include filing and reading, needs a lighting system that provides 120 footcandles per square foot. The room is 30 feet wide and 60 feet long with a 12 foot ceiling. Work desks are arranged in rows across the width of the room. The ceiling and walls are painted off-white and have a textured surface. The floor is light brown. A discussion with the owner indicates that the maintenance of the wall and ceiling surfaces should be good.

Surface reflectance is measured and found to be:

Ceiling 80%
Walls 50%
Floor 20%

An Acme luminaire No. 1224-4-40 is chosen. Manufacturer's data is given in Figure 6-8.

With all of the above information you can fill in the top part of the worksheet in Figure 6-9A.

Cavity ratios are calculated next. Use the formula given earlier in Step 5. When this is done, find the effective reflectance for the ceiling cavity and floor cavity in Table 6-7. Note that each cavity uses its own ratio and *surface* reflectance values.

Find the coefficient of utilization by applying the *effective* reflectances and the *room* cavity ratio in Table 6-8.

Finally, from the formula in Step 6 you find that 34 fixtures are needed.

The last step in the design is to locate the fixtures. The ratio between the spacing to mounting height is 1.4.

1.4 x 10 feet = 14 feet

The desks are in rows across the room width. Run the light fixtures the same way. Each fixture is four feet long. Seven placed end to end are 28 feet long. This fits the 30 foot room very nicely. But 34 fixtures divided by 7 rows comes out one fixture short of five rows. The owner doesn't want one short row of lights, so add one more fixture to make five rows of seven fixtures each.

The first and last row are located 6 feet from the wall. The other 3 rows are 12 feet apart, 2 feet less than the maximum spacing ratio (Figure 6-9B).

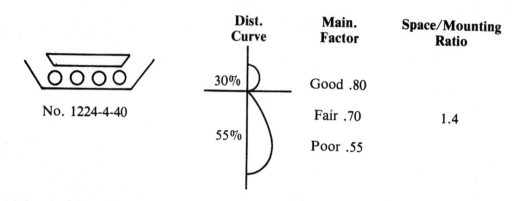

Coefficients Of Utilization												
Ceiling Cavity Eff. Reflec.	80			70			50			30		
Wall Reflec. (No. eff. reflec.)	50	30	10	50	30	10	50	30	10	50	30	10
Floor Cavity Eff. Reflec.	20											
Room Cavity Ratio												
0	.91	.91	.91	.86	.86	.86	.68	.68	.68	.61	.61	.61
1	.81	.78	.76	.77	.74	.72	.69	.67	.66	.56	.54	.54
2	.72	.68	.64	.69	.65	.61	.62	.59	.57	.51	.49	.47
3	.65	.59	.55	.62	.57	.53	.56	.52	.49	.46	.44	.42
4	.58	.52	.48	.56	.50	.46	.51	.46	.43	.46	.43	.40
5	.52	.46	.41	.50	.44	.40	.46	.41	.37	.42	.38	.35

Fixture Size:	12" x 48"
No. of Lamps:	4
Lamp Size:	40 watt
Lumens/Lamp:	2900
Trim Color:	White
Diffuser:	Clear or opaque prismatic plastic
Total Watts:	1600
Mounting:	Surface or suspended

Acme Fixture Specifications
Figure 6-8

No. *1*

Project *Example 1*   Date _____

Room *Inspec. Lab*   Footcandles/Sq. Ft. *120*

Room Dimensions	**Reflectance**	**Fixture Data**
Length *60 ft*	Wall *50* %	Mfr. *Acme*
Width *30 ft*	Ceiling *80* %	Cat. No. *1224 - 4 - 40*
Floor area *1800 sq ft.*	Floor *20* %	Lamps/Fix. *4 - 40 W*
Ceiling height *12 ft.*		Lumens/Lamp *2900*
Fixture height *10 ft.*		Coeff. of Utilization (CU) *.69*
		Maintenance Factor (MF) *.8*

Cavity Data

Room cavity
- Height *7.5*
- Ratio *1.9*

Ceiling cavity
- Height *2*
- Ratio *.5*
- Eff. Reflectance *.72*

Floor cavity
- Height *2.5*
- Ratio *.6*
- Eff. Reflectance *.19*

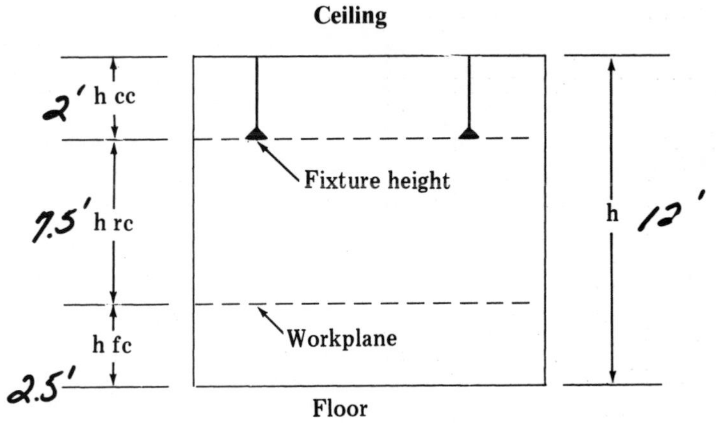

Ceiling

2' h cc

7.5' h rc Fixture height

h 12'

2.5' h fc Workplane

Floor

$$\frac{5 \times \text{cavity height} \times (\text{length} + \text{width})}{\text{Length} \times \text{width}} = \text{Cavity Ratio}$$

$$\frac{5 \times 7.5 \times (60 + 30)}{60 \times 30} = 1.9 = \text{RCR} \qquad \frac{5 \times 2 \times (60 + 30)}{60 \times 30} = .5 = \text{CCR}$$

$$\frac{5 \times 2.5 \times (60 + 30)}{30 \times 60} = .6 = \text{FCR}$$

$$\frac{\text{Floor area} \times \text{footcandles}}{\text{Lamps/Fix.} \times \text{Lumens/Lamp} \times \text{CU} \times \text{MF}} = \text{No. of Fixtures}$$

$$\frac{1800 \times 120}{ \times \times } = \frac{216000}{6403} = 33.7 = 34$$

$$\frac{\text{No. of Fixtures} \times \text{Lamps/Fix.} \times \text{Lumens/Lamp} \times \text{CU} \times \text{MF}}{\text{Floor area}} = \text{Footcandles}$$

$$\frac{ \times \times \times }{} =$$

Example 1 — Worksheet
Figure 6-9A

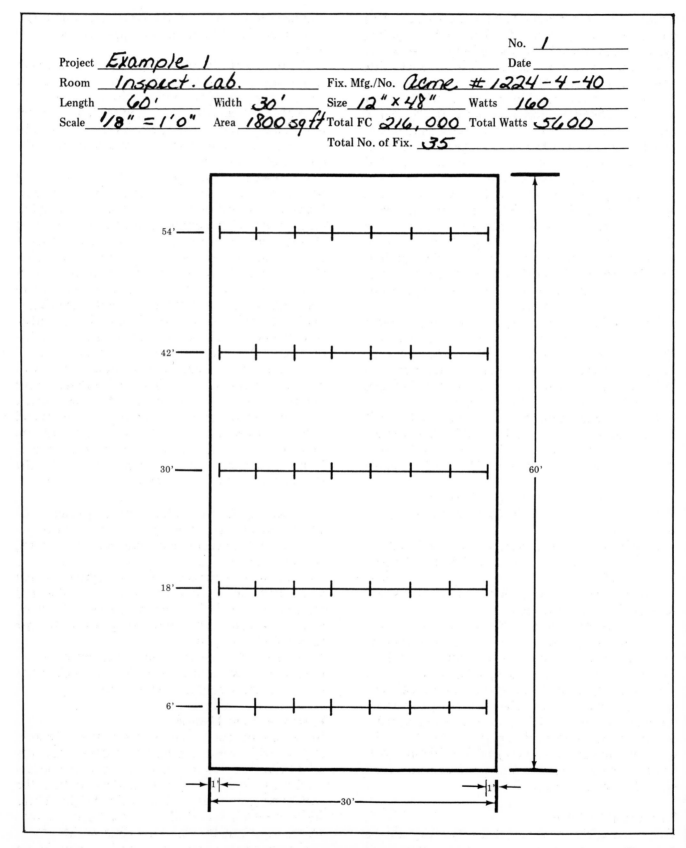

Example 1 — Fixture Location Worksheet
Figure 6-9B

Example 2 - Automobile Service Garage A service garage measuring 60 by 100 feet needs a lighting system. Tasks include all types of auto repair work. The ceiling is 30 feet high and has exposed joists and ductwork. Its reflectance is estimated at 60%. The walls are concrete block with a reflectance of 50%. The concrete floor is estimated at 20%. Since most of the repair work is done with supplementary lighting, only a fair level of general illumination is needed. The fixtures will be turned on and left on for long periods of time.

You judge that 50 footcandles per square foot will be enough light. The maintenance factor is fair. A Superior Hi-Bay luminaire is chosen. See Table 6-10. This fixture uses a 400-watt metal-halide lamp. It has a wide pattern of distribution. The fixtures will be mounted on the joists, 3 feet below the ceiling. The workplane for the garage is about 4 feet above the floor.

Figure 6-11A is the worksheet for this job. Note that in this example the floor cavity ratio is 0.5. That falls between the 0.4 and 0.6 cavity ratios for cavities with a 20% base and 50% wall reflectance in Table 6-7. Since the difference is only between 19 and 20, a value of 20% is chosen.

To use the manufacturer's CU table, round off the Ceiling Cavity Eff. Ref. to 60%. Then interpolate to get the correct CU. Since 60% CC Eff. Ref. is midway between the 50% and 70% table values:

1. RCR = 3 CC Eff. Ref. = 50% Wall Ref. = 50% FC Eff. Ref. = 20% Coefficient of Utilization = 0.66

2. RCR = 3 CC Eff. Ref. = 70% Wall Ref. = 50% FC Eff. Ref. = 20% Coefficient of Utilization = 0.68

Interpolation is easy, since the numbers are separated by only two digits:

Correct coefficient of utilization = 0.67

The final step is to locate the fixtures. A total of 18 are needed. The space to mounting ratio is 1. This gives a maximum spacing of 27 feet.

Since the level of illumination isn't critical, you can delete two fixtures. The remaining 16 can be arranged in 4 rows of 4 fixtures each. Fixtures spaced across the width could begin 7.5 feet from the wall and then be placed every 15 feet. Fixture rows would begin 12.5 feet from the end and be placed every 25 feet. See Figure 6-11B.

Residential Lighting

The zonal cavity method calculation may not be necessary when planning residential lighting. In most cases, the manufacturer's designation of the type of luminaire (kitchen, bathroom, etc.) is all you need to know. Still, the same rules that apply to commercial lighting design apply in residential work.

Good general illumination, proper supplementary lighting over work areas and elimination of glare are the primary considerations. Appearance both of the illumination itself and the luminaire are important in commercial applications. They can be just as important as when you're dealing with the owners of a new home.

Fluorescent luminaires in the kitchen and bathroom are common on most residential jobs. Other diffuse lighting will probably be incandescent. There are hundreds of designs and styles to choose from. (Figure 6-12). You may want to leave a good catalog with the owner and have him pick the fixtures he likes most. Some electrical contractors send the owner to a fixture wholesaler where many fixtures are on display. When the selection is made, you order the fixture and install it. In either case, there's extra profit for you because of the difference between your cost and the retail price.

On some jobs the lighting fixtures will be provided by an interior designer. If you have an indecisive and particular owner, this is a good choice because it relieves you of having to deal with the whims and personal tastes of the owner. The extra profit you make on the fixture markup isn't worth the problems it may cause. In this case, you specify only the wattage and type of mounting.

Energy Conservation and Lighting

The steady increase in energy costs has created genuine interest in energy conservation. The lighting system uses most electricity consumed in residential and commercial applications. The lighting system you install should be as efficient as possible and must comply with the code in any case. In many areas new codes now allow only 1½ watts per square foot for lighting.

You should be aware of these new code restrictions and know about the new energy-efficient fixtures that are available. (Figure 6-13.)

Energy-efficient Fixtures

In new construction, lighting calculations for low-energy fixtures are just like calculations for any other type of fixture. Use the manufacturer's specifications and recommendations to find the correct type, number, and mounting height. There are, however, a few pitfalls that you have to avoid. This is especially true in remodeling work.

The following suggestions will help you design an energy-saving lighting system that gives the best possible light at lowest cost. Please don't get the

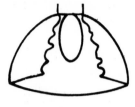

No. 36-12-1-400

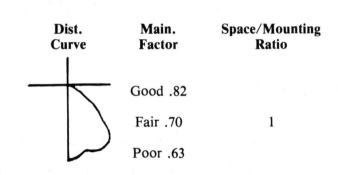

Dist. Curve	Main. Factor	Space/Mounting Ratio
	Good .82	
	Fair .70	1
	Poor .63	

Coefficients Of Utilization												
Ceiling Cavity Eff. Reflec.	80			70			50			30		
Wall Reflec. (No Eff. Reflec.)	50	30	10	50	30	10	50	30	10	50	30	10
Floor Cavity Eff. Reflec.	20											
Room Cavity Ratio												
0	.91	.91	.90	.89	.89	.88	.81	.81	.81	.78	.78	.78
1	.84	.82	.80	.82	.80	.78	.79	.78	.76	.76	.74	.73
2	.77	.73	.70	.76	.72	.71	.74	.73	.68	.70	.68	.67
3	.71	.66	.63	.68	.64	.60	.66	.63	.60	.64	.61	.58
4	.63	.58	.53	.62	.58	.54	.60	.56	.52	.58	.55	.52
5	.57	.51	.47	.56	.51	.47	.55	.50	.48	.53	.48	.46

Superior Hi-Bay	
Fixture Size:	14" dia.
No. of Lamps:	1
Lamp Size:	400 watt metal-halide clear
Lumens/Lamp:	36,000
Trim Color:	Black Enamel
Diffuser:	Clear glass
Total Watts:	400
Mounting:	Surface or suspended

Superior Fixture Specifications
Table 6-10

No. *1*

Project *Example 2*   Date

Room *Service Garage*   Footcandles/Sq. Ft. *50*

Room Dimensions
Length *100*
Width *60*
Floor area *6000*
Ceiling height *30*
Fixture height *27*

Reflectance
Wall *50* %
Ceiling *60* %
Floor *20* %

Fixture Data
Mfr. *Superior*
Cat. No. *36 - 12 - 1 - 400*
Lamps/Fix. *1*
Lumens/Lamp *36000*
Coeff. of Utilization (CU) *.67*
Maintenance Factor (MF) *.7*

Cavity Data

Room cavity
Height *23*
Ratio *3*

Ceiling cavity
Height *3*
Ratio *.4*
Eff. Reflectance *57%*

Floor cavity
Height *4*
Ratio *.5*
Eff. Reflectance *20%*

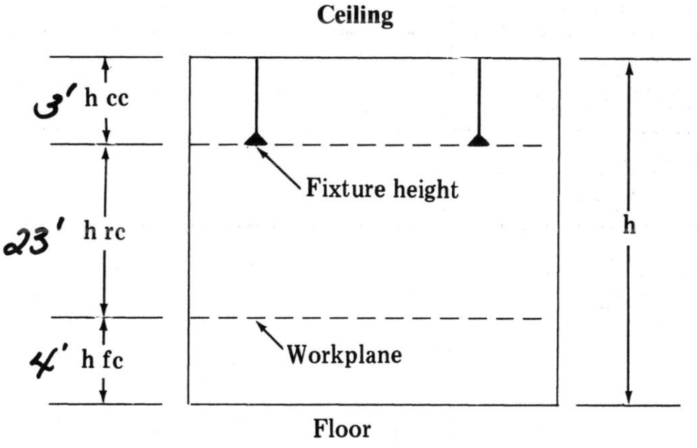

Ceiling

3' h cc

23' h rc

4' h fc

Fixture height

Workplane

h

Floor

$$\frac{5 \times \text{cavity height} \times (\text{length} + \text{width})}{\text{Length} \times \text{width}} = \text{Cavity Ratio}$$

$$\frac{5 \times 23 \times (100 + 60)}{100 \times 60} = 3 = \text{RCR} \qquad \frac{5 \times 3 \times (100+60)}{100 \times 60} = .4 = \text{CCR}$$

$$\frac{5 \times 4 \times (100+60)}{100 \times 60} = .5 = \text{FCR}$$

$$\frac{\text{Floor area} \times \text{footcandles}}{\text{Lamps/Fix.} \times \text{Lumens/Lamp} \times \text{CU} \times \text{MF}} = \text{No. of Fixtures}$$

$$\frac{6000 \times 50}{1 \times 36000 \times .67 \times .7} = \frac{300000}{16884} = 17.7 = 18$$

$$\frac{\text{No. of Fixtures} \times \text{Lamps/Fix.} \times \text{Lumens/Lamp} \times \text{CU} \times \text{MF}}{\text{Floor area}} = \text{Footcandles}$$

$$\frac{ \times \times \times }{} =$$

Example 2 — Worksheet
Figure 6-11A

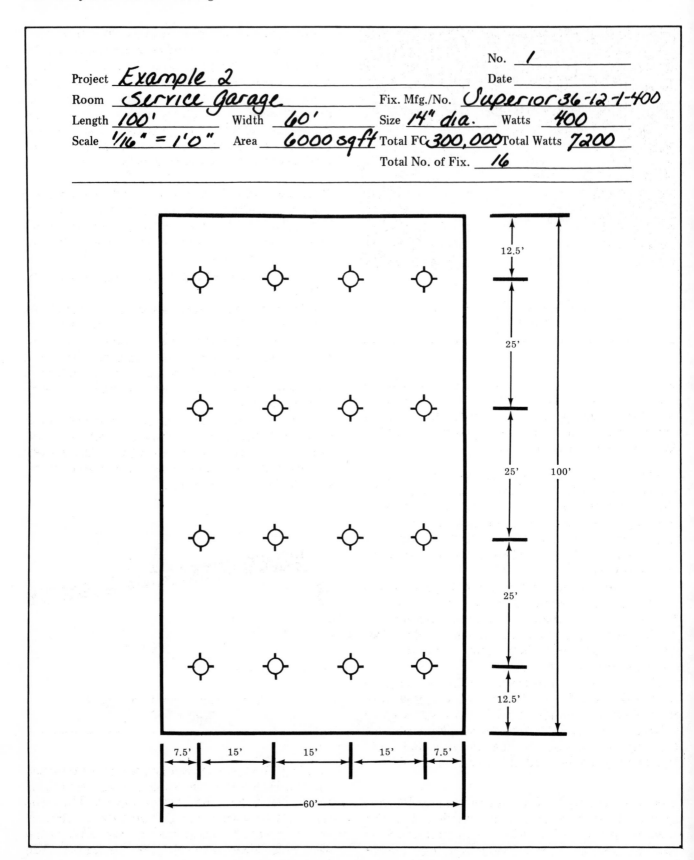

Example 2 — Fixture Location Worksheet
Figure 6-11B

Courtesy: Thomas Industries

A Dual-Function Residential Luminaire
Figure 6-12

impression that you should avoid energy-efficient fixtures. Just the opposite is true. When at all possible, recommend low-energy fixtures. Energy conservation benefits all of us. But you do need to know some basic ground rules:

Productivity - Productivity in work areas is directly related to the quality of light. Generally, the more difficult the task—reading small print, working with small objects—the more the quality and quantity of the light should increase. A decrease in either of these beyond certain limits will reduce productivity.

You save electricity while improving productivity by using *selective lighting*. With selective lighting, each functional area in the building or work area is lighted to the quality and level needed. Overall background illumination can then be reduced. See Table 6-14.

Here's the key to good selective lighting design in a remodeling job: scrutinize each existing work area. Over a building's useful life, tasks needing different light levels become mixed. The remodeled building should separate those tasks so that selective lighting can be used effectively.

Heat Loss and Gain - The problem of heat loss and gain is most important in remodeling. Lighting gives off plenty of heat which has an effect on the heating and air conditioning systems of the building. During the design phase, the architect or mechanical engineer will take this heat load into consideration.

But many of the new, energy-efficient fixtures operate at lower temperatures. If several newer fixtures are used to replace older fixtures, less heat is available. The opposite may also occur when more lighting is installed. Before making any major fixture changes in a lighting system—new or existing— consider the change in heat. You may need engineering assistance for this. At the end of this section you'll find sources of help.

Retail Sales and Illumination - Lighting has a big impact on sales in commercial applications. Too little light or a lamp the wrong color—although saving energy—can reduce sales and create an unhappy store owner.

The following summary of fixtures may be helpful:

1. Incandescent fixtures are cheap to install and have good color rendition. They are good for clothing and furniture stores, for example, because the merchandise looks to the customer the way it would in his home. However, the incandescent bulb has a relatively short life and uses high amounts of energy.

2. Fluorescent fixtures use less energy and have a longer bulb life. But they're more expensive to install. The light gives excellent illumination. However, it has a cooler color which may not be suitable for some locations.

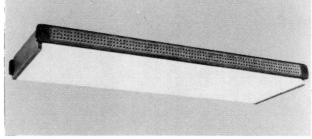

Courtesy: Thomas Industries

A Low-Energy Fluorescent Fixture
Figure 6-13

3. High Intensity Discharge (HID) fixtures (mercury vapor, metal halide, sodium vapor) are expensive to install but cost least to operate. The bulbs last a relatively long time. *Certain types* render color very close to the incandescent bulb. HID's are a wise choice where they will be on for four or more hours a day.

Tables 6-15 and 6-16 give more information about the different lighting types. If you're unsure

Circulation areas between work stations.	20 footcandles (fc.)
Background beyond tasks at circulation areas.	10 fc.
Waiting rooms and lounge areas.	10-15 fc.
Conference tables.	30 fc. with 10 fc for background lighting.
Secretarial desks.	50 fc. with auxiliary localized (lamp task lighting directed at paper holder (for typing) as needed. 60 fc. in secretarial pools.
Area over open drawers of filing cabinets.	30 fc.
Courtrooms and auditoriums.	30 fc.
Kitchens.	Nonuniform lighting with an average of 50 fc.
Cafeterias.	20 fc.
Snack bars.	20 fc.
Testing laboratories.	As required by the task, but background not to exceed 3:1 ratio footcandles.
Computer rooms.	As required by the task. Consider two levels, one-half and full. In computer areas, reduce general overall lighting levels to 30 fc and increase task lighting for areas critical for input. Too-high a level of general lighting makes reading self-illuminated indicators difficult.
Drafting rooms.	Full-time, 80 fc. at work stations. Part-time, 60 fc. at work stations.
Accounting offices.	80 fc. at work stations.

Suggested Lighting Levels
Table 6-14

about a specific installation, check with a lighting engineer or the manufacturer's rep. When remodeling, experiment with just one or two fixtures in a small area of the building. If the owner and occupants are satisfied with the appearance, then make a major change.

Energy-efficient Switching
Electricity can be saved by improved switching design. Light switches cost very little. Look for opportunities to install extra switches so unused fixtures can be off while needed fixtures stay on. The following methods apply to both new work and remodeling.

Zone Switching In the past it was common practice to wire an entire lighting system to a single switch. An entire floor or room of a building had to be switched on at one time. We know now that it's wiser to install enough switches so only the lights needed at any one time can be switched on. Large offices, warehouses and storage areas are typical locations which benefit from zone switching.

Type	Lumens Per Watt	Advantages	Disadvantages
Incandescent		Inexpensive	Low efficiency
general service	21	Good color rendition	Short lifetime
extended service	15	No auxiliary equipment	Sensitive to voltage input
tungsten-halogen	22	Easy to install	High surface brightness
		Instant startup	Burns "hot"
		Easy to dim, flash, and control	
Fluorescent	80	Moderate efficiency	Requires auxiliary equipment
		Longer life	No instantaneous startup
		Low surface brightness	Temperature sensitive
		Burns "cool"	Lamp lifetime depends on starts per day (more starts = shorter life)
High-intensity Discharge		Higher efficiency at higher wattages	No instant startup
Mercury			
Clear	46	Long lamp life	Poor color rendition
Phosphorous	52	Good optical control	Requires auxiliary equipment
Metal Halide			
Clear	70	Same as mercury	Requires auxiliary equipment
Phosphorous	90	Good color rendition	No instant startup
High Pressure Sodium	120-130	Highest efficiency Long lamp life	Same as mercury
Low Pressure Sodium	180	Minimal lumen depreciation	

Characteristics of Lighting Types
Table 6-15

Multiple Switching Three and four-way switches, giving light control from more than one location, are also effective in saving energy. If people have to double back through a dark room to throw a switch, they'll usually leave the lights on. Multiple switching eliminates this problem. Chapter 7 has wiring diagrams for multiple switching.

Retail Display Lighting For retail areas, display lighting should be switched independently from overhead lights. That way it can be turned off when not needed. When the display lighting is on, the opposite situation may work. Overhead lighting above the display area may be turned off or reduced.

When designing display lighting, try to group several display areas to maximize the use of one or two display lighting fixtures.

Time Switches and Photo-electric Switches Display lighting, exterior building lights and sign lights are often best switched by an automatic on-off control. These controls eliminate human error and permit control when the building is unoccupied.

Time switches come in a wide variety of models. For restrooms, storerooms or other areas that have short occupancy, a time switch will turn the lights off if the user forgets.

Time switches also give lighting control when a building is unoccupied. For example, studies show that vehicle traffic drops off sharply at midnight. Signs and building lights left on after hours for their advertising value can be turned off with a timer when they are no longer effective.

Photo-electric switches can be used to control artificial light in building interiors. For example, a large office can have its light rows wired to a dimmer-type photo-electric switch. Early in the

Existing	Replacement
S t o r e Sixty-six 500-W incandescent down lights	Twenty 400-W metal halide down lights
Total kWh/yr. = 122,100	Total kilowatt hours per year = 34,040
Total savings = 88,060 kWh/yr. = 72% savings	
Seventeen 1000-W incandescent pendant fixtures	Fourteen 400-W metal halide pendant fixtures
Total kWh/yr. = 62,900	Total kWh/yr = 24,050
Total savings = 38,850 kWh/yr. = 62% savings	
Parking Area Twelve 1500-W tungsten halogen lamp floodlights	Four 1000-W metal halide cluster
Total kWh/yr = 27,000	Total kWh/yr = 6480
Total savings = 20,520 kWh/yr. = 76% savings	
Sign Two 1500-W tungsten halogen floodlights	Two 400-W metal halide floodlights
Total kWh/yr. = 12,000	Total kWh/yr. = 3680
Total savings = 8320 kWh/yr = 69% savings	
Security Six 500-W tungsten halogen floodlights	Six 150-W high-pressure sodium lights
Total kWh/yr = 12,000	Total kWh/yr = 3920
Total savings = 8080 kWh/yr = 67% savings	
Supermarket Two hundred fifty 4-lamp 40-W strip fluorescent fixtures	Seventy 400-W high-pressure sodium down lights (where change in color rendition is acceptable)
Total kWh/yr = 146,250	Total kWh/yr = 84,000
Total savings = 62,250 kWh/yr. = 43% savings	

Examples of Lighting Conservation. kWh/yr. = Kilowatt Hour Per Year
Table 6-16

day, all the lights will be on. As more daylight comes in through the windows, the switch will decrease the artificial light while still keeping the illumination in balance. Later in the day, the process is reversed.

Photo-electric switches also work well for exterior security lighting and parking areas. See Figure 6-17. The fixtures should be HID type. Provide enough light but avoid waste. Put the best light near high-risk areas such as doors, windows and pedestrian walks. Reduce lighting in low-risk areas such as against a building wall.

Exterior Fixture—Low-Pressure Sodium With Photo-Electric Switch
Figure 6-17

Dimmers Many rooms in a commercial building are used for a variety of purposes, each with a different lighting need. Dimmers are a good solution when adjustable lighting is possible. However, keep in mind that dimmers are only effective if someone controls them properly. If possible, reduce light levels permanently through lower-wattage fixtures or use zone switching. If this isn't possible, dimmers are the next best choice.

Dimmers work especially well with incandescent fixtures in single-fixture rooms. They are also available for fluorescent lights, but multi-level ballasts must be used. This increases the installation cost.

Duplicate Switching Many multiple lamp fixtures are wired so that half of the lamps in each fixture can be switched separately from the other half. Each room then has two light switches. If only one is on, the lighting level is only half intensity. That's enough for many uses. Even if you don't use multiple lamp fixtures, consider switching half of all fixtures in a room on one circuit and half on another. This is appropriate in nearly all industrial and office buildings and will apply in other construction where room size is large.

Some codes now require that industrial and commercial buildings have duplicate switching so lighting levels can be reduced by half. It boosts installation costs by several percent, but will save electricity and may prove invaluable in an electrical shortage emergency.

Payback

You'll meet the builder or owner who wants to know how much can be saved by installing an energy-saving lighting system. Basically, dollar savings are found by dividing the cost of the improvement by the savings that result.

For example, fluorescent lights give three to four times more lumens per watt than incandescent lights. They're clearly cost-effective if they operate at least 40 hours per week.

Assume that the owner of a 120 square foot office wants to know how soon his investment will be paid back if he replaces his incandescent fixtures with fluorescent. Here's your problem:

Light Level   = 100 ft-candles
Ceiling Area   = 120 sq/ft.
Ceiling Height = 10 ft.

Current number of fixtures:
  8 - 100 watt incandescent
Equivalent replacements:
  2 - 2 lamp 40 watt fluorescent fixtures
  (92 watts with ballast)

Load:
  Incandescent - 800 watts (6.6 watts per sq./ft.)
  Fluorescent - 184 watts (1.5 watts per sq./ft.)

Annual Cost: (@5¢/KWH - 8 hour day, 260 workdays)

Incandescent	$83.20
Fluorescent	19.10
Annual Savings	**$64.10**

Assuming an average cost per fixture of $75, the two new fluorescent fixtures will be $150.

$$\frac{\$150.00}{\$64.10} = 2.3 \text{ years payback}$$

In some cases, you'll need more information in the payback calculation. Some areas allow tax credits for energy-saving improvements. If bulb and maintenance costs are substantially lower, it may be worthwhile to consider this as well.

Case Study
Here's an example of how energy-efficiency was combined with good illumination to meet a particular requirement. The design was done by a professional firm, but the solution illustrates what any electrical contractor could do on any size commercial job with modern lighting fixtures and a little imagination.

The owners of a new department store in California wanted to limit total electricity input to 1.6 watts per square foot—about half the wattage normally used in department stores. Based on their experience at similar stores, they expected savings for the 150,000 square foot building to be about $7500 annually if the designer stayed within the energy budget. The designer's challenge was to provide good light of the right color by combining daylight with energy-efficient fixtures.

The designer cut electrical input by taking advantage of the natural light that was available and using high intensity discharge fixtures for general illumination. Here's how he did it.

In one area of the store, the need for daytime artificial light was reduced by installing a white fiberglass canopy as part of the roof. The canopy admits about 500 footcandles of light during daylight hours. It also attracts shoppers to the store with its unique design. Its underside is specially coated for high reflectance.

At night and on overcast days, ceiling-hung fixtures are directed at the canopy to give indirect illumination. The fixtures are a combination of 1000 watt metal-halide and low-pressure sodium lamps. Metal halide fixtures alone would give merchandise a harsh appearance because of their brightness. The warm amber of the low-pressure sodium fixtures softens and balances the light to display the merchandise in its natural color.

Normally, metal-halide and low-pressure sodium fixtures are used outdoors. But with five times the efficiency and ten times the lamp life of the traditional quartz halogen lamps used in department stores, they were very appealing. The store owners report that color rendition equals or surpasses that used in stores with more traditional lighting.

Where product visibility was not important, 2 foot square parabolic fluorescent fixtures were used. In special display areas, fluorescent lamps were used for background illumination. Merchandise was highlighted by low-voltage track lighting.

Need More Information
A good source for more information of energy-efficient lighting is the National Lighting Bureau. They offer a catalog of publications.

National Lighting Bureau
2101 L Street, NW/Suite 300
Washington, DC 20037

Consulting engineers who specialize in lighting can be found by contacting your local chapter of the Illuminating Engineering Society of North America. If you don't have a local chapter, write the national headquarters at:

Illumination Engineering Society
United Engineering Center
345 East 47th St.
New York, NY 10017

The following association stays abreast of recent innovations in lighting system technology:

National Association of Electrical Distributors
600 Summer Street
Stamford, CT 06901

Another source for information is the fixture manufacturers. They're listed in your phone book.

Now that we've looked at design of commercial lighting systems in detail, we can go on to designing the electrical circuits.

Branch Circuits
A *branch circuit* is wiring run between the circuit breaker or fuse and individual loads. It includes the breaker or fuse and the outlet or fixture wiring. Branch circuits are almost always the last part of the distribution system between the main or feeder equipment and the system loads.

Branch circuits supply lighting and appliance loads. Usually they supply both, but may supply only one or the other. They also supply motor loads in combination with lighting or appliances or alone in the case of a larger motor.

Item	Amps	Wires	Volts
Range (up to 12 Kw rating)	50	3	115/230
Combination washer/dryer or	40	3	115/230
Automatic washer	20	2	115
Electric dryer	30	3	115/230
Fuel-fired furnace	15/20	2	115
Dishwasher	20	3	115/230
Water heater	30	3	115/230
Water pump	20	2	115

Individual Branch Circuits
Typical Dwelling
Table 6-18

Branch circuits are classified by the maximum rating or setting of the overcurrent device as allowed by the NEC. Circuits of 15, 20, 30, 40 and 50 amps are common. Sometimes separate circuits of higher ampacities are needed. In this case, the ampere rating or setting of the overcurrent device determines the circuit classification.

Types of Branch Circuits
There are 3 basic types of branch circuits.

General Purpose - A general purpose branch circuit supplies 115 volt power to lighting and convenience outlets in light commercial applications and dwellings. Bedrooms, living rooms, hallways, offices and lobbies are typical of rooms using general purpose branch circuits. These will be 15 or 20 amp circuits. Outlets supplied by these circuits should be divided as evenly as possible among the circuits planned. But the NEC doesn't permit more than 10 outlets on a 20 amp circuit or 7 on a 15 amp circuit. Check local codes too, since some may have stricter limits.

It's good practice not to put all outlets on any one floor or area on the same circuit. This keeps some lights on when a circuit overloads.

Appliance Circuit - Appliance circuits are 20 amp branch circuits that also operate at 115 volts. They're used to meet NEC rules on the minimum number and size of circuits in kitchens and laundry rooms in dwellings. Two or more circuits are required in the kitchen. One or more are required in the laundry room area.

These circuits can't have any permanently connected *lighting* outlets unless they are part of an appliance. They can't supply receptacles outside the room they are intended to serve. But kitchen appliance circuits can supply receptacles in a pantry, breakfast room or dining room.

Individual Circuit - An individual branch circuit supplies only one appliance. Items such as ranges, hot water heaters and electric heating units all will have dedicated circuits. Many of these draw 230 volts. Some may require only 115 volts. Sometimes a separate, fused or switched equipment connection is required. Table 6-18 lists the individual circuits you'll find in most dwellings.

As a general rule, individual circuits should be used for the following:
1. All motors of 1/2 hp or larger. (See Motor Section in this chapter for details.)
2. All stationary appliances rated over 1500 watts.
3. All equipment requiring continuous service, such as food freezers, heating systems and water pumps.
4. All 210-240 volt appliances.

Loads for Branch Circuits
The load allowed on a branch circuit depends on

Circuit Rating	15 A	20 A	30 A	40 A	50 A
Conductors* Min. Size Copper	No. 14	12	10	8	6
Overcurrent Protection	15 A	20 A	30 A	40 A	50 A
Outlet Devices Lampholders Receptacle Rating	Any Type 15 A Max.	Any Type 15/20 A	Heavy Duty 30 A	Heavy Duty 40/50 A	Heavy Duty 50 A
Maximum Load	15 A	20 A	30 A	40 A	50 A

* Type T, TW, RH, RUH, RHW, RHH, THHN, THW, THWN, XHHW, FEP, FEPB, RUW and SA in raceway or cable.

Branch Circuit Requirements
Table 6-19

the circuit rating. The *circuit rating* depends on three things:

1. Conductor size.
2. Overcurrent protection device size.
3. Receptacle or lampholder rating.

For example, a 20 amp circuit must use a minimum size of No. 12 wire. The fuse or breaker must be rated at 20 amps. The lampholders—if any are used—can be any type. Receptacles must be rated at 15 or 20 amps. Table 6-19 summarizes branch circuit requirements.

In no case can the load allowed exceed the circuit rating; a larger size circuit must be used or another circuit included in the plan.

An individual branch circuit can supply any load for which it is rated. For branch circuits supplying two or more outlets, the following limitations apply.

1. For *known loads,* the total load can't exceed 80% of the branch circuit rating if the circuit supplies:
   a. Motor loads 1/8 hp or less
   b. Portable appliances
   c. Continuous loads, such as display or overnight lighting.
2. No single appliance, fixed or portable, can exceed 80% of the branch circuit rating.
3. The total rating of fixed appliances can't be more than 50% of the branch circuit rating if lighting and portable appliances are also supplied.
4. For *unknown* loads, estimate 1.5 amps per outlet. Remember that the maximum load permitted is 80% of the circuit rating, whether the load is known or unknown.

Branch circuit sizing requirements for motors are explained later in this chapter.

If there is more than one air conditioning or refrigeration load on a circuit, refer to the section on Air Conditioning and Refrigeration in the NEC.

Branch Circuits Required

The rules above regulate the maximum load on a branch circuit. The minimum number of general purpose branch circuits is controlled by the square footage of the building and its use.

Table 6-20 lists general lighting loads required by the NEC for various types of buildings. For dwelling units, a minimum of 3 watts per square foot is needed. Total area is calculated using the outside dimensions of the building. Garages, porches, unfinished or unused spaces should not be included.

The 3 watt per square foot lighting load for dwellings takes into account any convenience receptacles on the circuit. No additional calculations are needed for general purpose circuits.

This 3 watt per square foot rule also applies to hotel and motel guest rooms and apartment units without cooking appliances. For office buildings and banks, add 1 watt per square foot to the lighting load to provide for convenience receptacles. For all other loads, the convenience outlet load can be calculated using 180 volt-amperes per outlet (1.5 amps).

Here's an example that will make this clearer to you. Suppose you want to find the minimum number of general purpose outlets needed for a

Occupany	Load/Sq. Ft. Watts
Auditoriums, churches	1
Banks, office buildings	3½*
Barber shops, beauty parlors	3
Clubs, court rooms	2
Dwelling units	3**
Garages, warehouses (storage)	½
Hospitals, restaurants	2
Hotels, motels, apartment houses without tenant cooking	2**
Schools, stores	2

* Add 1 w/sq. ft. for general receptacles when actual number is not known.

**Includes both lighting and general purpose convenience receptacles for 20 amp or less circuits. No additional load calculations are needed. NEC Section 220-A Small Appliance circuits must be added.

For additional lighting loads, consult NEC Section 220.

General Lighting Loads
Table 6-20

2,000 square foot dwelling:

2,000 square feet times 3 watts per square foot = 6,000 watts

If we assume each circuit is to be 20 amps, the formula Watts = Volts x Amps (P = E x I) can be used to find:

W = 115V x 20A = 2300 watts per circuit

$$\frac{6,000 \text{ total watts}}{2,300 \text{ watts/circuit}} = 2.6 \text{ circuits}$$

Rounding off 2.6, we find 3 general purpose circuits will meet code requirements.

If 15 amp circuits are used:

W = 115 x 15 = 1725 watts/circuit

$$\frac{6,000 \text{ total watts}}{1,725 \text{ watts/circuit}} = 3.5 = 4 \text{ circuits needed.}$$

It is also common practice to calculate general purpose branch circuits on the following basis:

1. One 20 amp 115 volt circuit for each 500 feet of floor space. Or;
2. One 15 amp 115 volt circuit for each 375 feet of floor space.

This method yields a higher number of circuits and more wattage. It's the preferred method when designing modern electrical circuitry.

Applying this rule on the same 2,000 square foot dwelling with 20 amp circuits:

$$\frac{2,000 \text{ total square foot}}{500 \text{ square foot/circuit}} = 4 \text{ circuits}$$

Since each circuit is rated at 2300 watts, total wattage will be 9200 watts rather than the 6000 watts found using the NEC minimum of 3 watts per square foot of floor.

Numbering Branch Circuits

When you've calculated the number of branch circuits needed, give each a number. Put this number next to each outlet, switch or appliance on the plan to show how the building is circuited. Circuits are titled and numbered using a panelboard sheet or schedule. See Figure 6-21.

The panelboard sheet will usually show two columns side by side for branch circuits. These columns show how the circuits are to be located on the panel. Location is important because it guarantees that the total circuit load will be balanced between the hot wires of the service. An unbalanced load is inefficient and can cause damage to the system. Figure 6-22 shows a 200 amp panel and the circuiting for a typical dwelling.

The panelboard sheet will also show total wattage of the system and will have columns for each circuit's amperage rating and the number of poles required.

The term *pole* refers to the number of hot wires which are connected to the breaker. Circuits that operate at 115 volts and use one hot wire and a neutral will use a single-pole breaker. Circuits that use two hot wires will need a double-pole breaker. A double-pole breaker will take up twice as much physical space in the box. It will also use higher amperage and watts. This is indicated by using two spaces on the panel sheet and using both numbers as the circuit number.

The finished branch circuit design should include spares in the panel for future use.

Feeder and Service Loads

Service can be defined as the conductors and equipment that deliver power from the utility's supply lines to the wiring system of the building. *Feeders* are the conductors between the service and the final branch circuit overcurrent device.

The difference between a feeder and a service conductor sometimes depends on the use. For example, the wires that enter a dwelling as the service wires may continue from the meter through the main disconnect into the branch circuit panel. Strictly speaking, the wires passing between the main disconnect and the panel are feeders. However, since the load they service is identical to the service load, there is no real difference between them. In this case, only the service load has to be calculated.

Suppose a building has a workshop, offices and a lighted display room—three unique uses spread out over a large building. A big single-service system would be costly to install and would necessarily have large voltage drops to the loads. It may be more convenient and efficient to give each area a separate distribution sub-panel. The wires between the main service panel and the separate supply panels would be feeder circuits. Their load would be different from the service load.

Service conductors and equipment are the main supply. They control power entering a building or area. With some exceptions, the NEC does not allow more than one service for any building or area. Feeder circuits, however, may distribute power to sub-panels in several parts of the building.

Feeder circuits are common on farms, where they serve separate buildings. On a remodeling job, a feeder circuit is often installed to serve the building addition. Multi-family dwellings use feeder circuits to carry power from the service panel to each dwelling unit.

You have to calculate service and feeder loads to select the right size wire and equipment. The same basic procedure is used for both service and feeder circuits: Find or estimate the loads. Then apply the demand factor as explained in the next section. Finally, divide the wattage calculated by the system voltage. This gives the ampacity needed in the wire and equipment.

Demand Factor

The *demand factor* is the ratio of the maximum demand of a system or part of a system to the total load connected. This just makes sense. You'll never have all the lights and appliances in a system on at the same time. There's no need to design the system as though all lights and appliances were going to be in use at once. The demand factor tells us what percentage of total possible use we have to plan for. This way the NEC lets us install wiring smaller than needed if the system really were operated at maximum demand.

Demand factors are given as percentages of a load or part of a load, listed in watts. These percentages have a wide range. Generally the determination depends on the following:

1. The type of building in which the circuit, appliance or device is located.

2. The type of appliance, circuit or device.

3. The number of appliances or devices connected to the feeder or circuit.

For example, in a dwelling, the NEC requires a demand factor of 100% for the first 3,000 watts of lighting load. The remainder is calculated at 35%, up to 120,000 watts. An electric clothes dryer located in the same dwelling would have a load of 5,000 watts or the nameplate rating, whichever is higher. Its demand factor must be computed at 100%. List the demand factor and wattage of all the general branch circuits and large appliances in the system to find the total load. See Table 6-23.

Panelboard Schedule

Panel A   **Type** SQD   **Volts** 240   **Phase** 1   **Wire** 3

Location Basement   **Amps** 200   **Mains** □ Lug ☒ Breaker

Flush □   Surface ☒

Branch Breakers

Item	Amp Rating	Pole	Cir. No.	Left	Right	Cir. No.	Pole	Amp Rating	Item
Range	50	2	1	12,000	5000	2	2	30	Dryer
—	—	1	3	—	—	4	1	—	—
Air Cond.	30	2	5	1400	5000	6	2	30	Water Heat.
—	—	1	7	—	—	8	1	—	—
Lighting	20	1	9	2300	1200	10	1	15	Dishwasher
"	20	1	11	2300	2300	12	1	20	Appl. Circuit
"	20	1	13	2300	2300	14	1	20	"
Furnace	15	1	15	600	2300	16	1	20	"
Elec. Heat	30	2	17	2100	2100	18	2	30	Elec. Heat
—	—	1	19	—	—	20	1	—	—
Elec. Heat	30	2	21	2100	2100	22	2	30	Elec. Heat
—	—	1	23		—	24	1	—	—
Spare			25			26			Spare
			27			28			
				25100	22300				Total Conn Load

Top ☒   Bottom □   **Equip. Rating** #000   **Feeder Size** 200 Amps

Panelboard Schedule
Figure 6-21

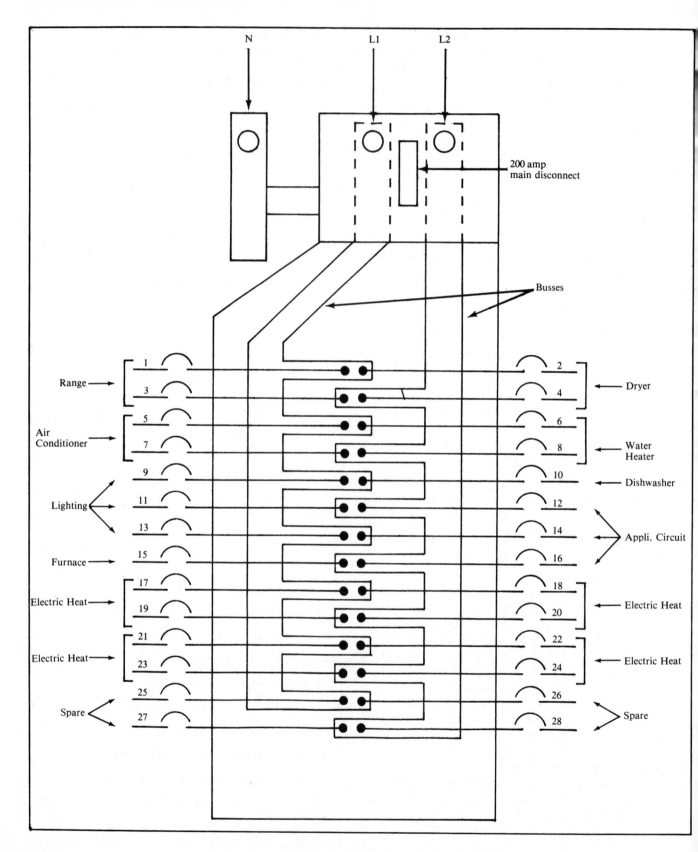

200 Amp Panel and Circuiting for a Typical Dwelling
Figure 6-22

Demand Factors — Dwelling Units

Type of Load	Portion of Load to Which DF applies (wattage) or Number of Appliances	Demand Factor (%)
Lighting load	First 3,000 or less Next 3,001 to 12,0000 Remainder over 120,000	100% 35% 25%
Small appliance/laundry circuits	4,500. Include in lighting load calculation	----
Electric dryer	1 - 4 units 5 units 6 units	100% 80% 70%
Electric space heating	Total connected load	100%
Air conditioning	Omit if less than electric heating. If more, consult NEC Sec. 440.	100%
Fixed appliance load. (Cannot include ranges, dryers, space heating and air conditioning.)	1 - 3 units 4 or more served by same feeder.	100% 75%
Ranges, ovens, fixed counter cooking units, and any other household cooking units rated 8,750 to 12,000 watts.	1 unit 2 units 3 units	No percent. Use 8000 W 11000 W (total) 14000 W (total)

For other loads, consult NEC Section 220.

Demand Factors — Dwelling Units
Table 6-23

When finding the demand, you can ignore one or more of several loads that won't be used at the same time. For example, in a dwelling it's unlikely that a room air conditioner will be used when a baseboard electric heater is running. In cases like this, it's acceptable to ignore the smaller of the two loads when making your calculations.

Other options are also available for calculating the demand. See Section 220 of the NEC and the examples in this chapter.

Estimating Appliance Loads
Use these allowances when estimating appliance loads unless the actual equipment rating is known. Then use the actual load.

Range - 12,000 watts

Water heater - 4,500 to 6,000 watts

Clothes dryer - 5,000 watts

Clothes washer - 600 watts

Dishwasher - 1,200 watts

Disposal - 400 watts

Food freezer (large) - 1,200 watts

Water pump - 1,000 watts

Bathroom heater - 1,500 watts

Ventilating fan (small) - 500 watts

Fuel-fired furnace - 600 watts

Room air conditioner - 1,400 watts

Portable microwave range - 1,450 watts

Electric Space Heating - Manufacturer's nameplate rating

Central Air Conditioning - Manufacturer's nameplate rating

Calculating Service and Feeder Loads
Single-family Dwellings The NEC requires a 100

amp, 3-wire service for any single-family dwelling with six or more 2-wire branch circuits or a load of 10 kw or larger. The ungrounded service conductors must not be smaller than No. 8 copper, or equivalent. Smaller systems can use lesser services down to a minimum of 60 amps but this is not recommended. Residential service of less than 100 amps allows only the essentials. There's little room for expansion.

Here's how to find the service entrance capacity:

Example 1 - A dwelling has 1660 square feet of floor area not including an unoccupied basement, unfinished attic and open porch. It has a 12 kw range, 4.5 kw water heater, 0.6 kw washer, 1.2 kw dishwasher, a 0.6 kw water pump and a 0.5 kw fuel-fired furnace.

Computed Load:

General Lighting Load, NEC 220-A
1660 sq. ft. x 3w/ft.   = 4980 W

Small Appliance Load, NEC 220-A
1500 W x 2 circuits   = 3000 W

Laundry Circuit, NEC 220-A
1500 W x 1 circuit   = 1500 W
                    Total   9480 W

Minimum Number of Branch Circuits:
General Lighting Load
NEC Section 220A

$$\frac{4980 \text{ W}}{115 \text{ V}} = \begin{array}{l} 43.3 \text{ amps} = \text{three 15 amp 2 wire} \\ \text{circuits or} \\ \text{two 20 amp and one 15 amp 2 wire} \\ \text{circuits} \end{array}$$

Small Appliance Load

NEC Section 220-A = two 20 amp 2 wire circuits

Laundry Load

NEC Section 220-A = one 20 amp 2 wire circuit

Demand Factors:

Lighting and Appliance Load, NEC 220-B
    First 3000 W at 100%   = 3000 W
    9480 - 3000 at 35%   = 2268 W
                    Total   5268 W

Range, NEC 220-B   = 8000 W

Fixed Equipment, NEC 220-B

Water Heater	=	4500 W
Dishwasher	=	1200 W
Washer	=	600 W
Water Pump	=	600 W
Furnace	=	500 W

    Total   7400 W at 75%* = 5550 W

*75% Demand factor applies for four or more fixed appliances. Three or less would use 100%.

Total Wattage:

Lighting and Appliance Load	= 5268 W
Range	= 8000 W
Fixed Equipment	= 5550 W

                    Total   18818 W

Service Capacity:

$$\frac{18818 \text{ W}}{230 \text{ V}} = 81.8 \text{ amps}$$

The ampacity of this service of 81.8 amps is within the 100-amp size required by the NEC for loads of 10,000 W or more. A 100-amp service will safely handle the planned load as well as fulfill NEC rules.

Example 2 - The NEC permits an optional calculation for dwellings which use a 100-amp service, provided some other conditions are met. (See Section 220-C.) For the same dwelling in Example 1, suppose the fuel-fired furnace is replaced by 8.5 kw of electric heat, separately controlled in six rooms. A 7-amp, 230-volt air conditioner is also added. For this system, the service capacity can be calculated:

General Lighting Load		
1660 sq. ft. at 3 W	=	4980 W
Small Appliance Load	=	3000 W
Laundry Load	=	1500 W
Range	=	12000 W
Water Heater	=	4500 W
Dishwasher	=	1200 W
Washer	=	600 W
Water Pump	=	600 W
House Heating	=	8500 W
Total	=	36880 W

First 10 kw at 100%	=	10000 W
Remainder at 40% (26880 x .40) =		10752 W
Total Wattage	=	20752 W

Service Capacity:

$$\frac{20752 \text{ W}}{230 \text{ V}} = 90 \text{ amps}$$

A 100 amp service can handle the electric heating and the air conditioner.

Note that in this example the full loads of the circuits and all appliances are given. The demand factor is calculated from their total. Appliance loads for the example were taken from appliance load estimates in this chapter.

But when using this method on the job, the NEC requires that the *nameplate* rating of all appliances be used.

Note also that the air conditioner was not figured in. This is because Section 220-B of the NEC permits the smaller of two unlike loads to be dropped from the service calculation. The 7 amp, 230 V air conditioner had a load of 1610 watts. This is less than the electric heat load. Since it is unlikely that both the heat and air conditioner would run at the same time, the air conditioner was dropped.

Multi-family Dwellings Figure service capacities for multi-family dwellings the same way you figure service for single-family residences. Once you have calculated each individual unit service, multiply by the number of like units in the building. Demand factors are applied to the various total loads. House loads, if there are any, are figured in. Then the total system capacity is calculated. An example will help clarify this.

Example 1- An apartment building with a total floor area of 11,200 square feet has 12 dwelling units. Each unit has 800 square feet for a total of 9,600 square feet of unit space. The remaining 1,600 square feet is divided among halls, a community laundry room, a manager's office, storage area and the mechanical room. These areas are served by a house panel. The house circuit also services central hot water heaters and a fuel-fired boiler for central heat.

The service runs from the main disconnect to a double bank of apartment meters. A feeder also runs between the main disconnect and the house panel.

Dwelling Unit Calculations:

Lighting Load - 800 sq. ft. x 3 W/sq. ft.	= 2400 W
Small Appliance Circuits	= 3000 W
Laundry Circuit	Omitted
Total General Load	5400 W

Minimum Number of Branch Circuits:

Lighting

$$\frac{2400 \text{ W}}{115 \text{ V}} = 21 \text{ amps} = \begin{array}{c} 2\text{-}15 \text{ amp 2 wire circuits} \\ \text{or} \\ 2\text{-}20 \text{ amp 2 wire circuits} \end{array}$$

Small Appliance = 2 - 20 amp 2 wire circuits

Demand Factor: General Load of 5400 W.

First 3000 W at 100%	= 3000 W
Remainder at 35% (5400 - 3000 = 2400)	= 840 W

Range Load:

12 kw Range	= 8000 W

Air Conditioner:

1100 W at 100%	= 1100 W

Fixed Appliances:

Disposal	500 W	
Dishwasher	1800 W	
Exhaust Fans	300 W	
Range Hood	450 W	
Total	3050 x 75% DF	= 2288 W
	Total Net Load	15228 W

Minimum Sub-feeder Size - Meter Bank to Units:

$$\frac{15228 \text{ W}}{230 \text{ V}} = 66 \text{ amps}$$

Minimum Size Feeder - Service Equipment to Meter Bank:

12 units

Total General Load:

Lighting and Small Appliances

12 x 5400 W = 64800 W

Demand Factor: General Load of 64800 W.

First 3000 W at 100%	= 3000 W
Remainder at 35% (64800 3000 = 61800)	= 21630 W

Range Load (NEC Table in Section 220B):

12 kw Ranges	= 27000 W

Fixed Appliance Load:

3050 W x 12 = 36600 W at 75% = 27450 W

Air Conditioners:

1100 W x 12 = 13200 W at 100% = 13200 W

Total 92280 W

Minimum Feeder Size - Main Disconnect to Meter Bank:

$$\frac{92280 \text{ W}}{230 \text{ V}} = 401 \text{ amps}$$

The NEC allows you to use a special calculation on three or more units in a multi-family dwelling (Section 220-C) if:

1. No single dwelling unit is supplied by more than one feeder.

2. Each dwelling unit is equipped with electric cooking equipment.

3. Each dwelling unit has electric space heating, air conditioning, or both.

The units in this example qualify. Each has an air conditioner. So it's O.K. to use the optional method. The advantage is that you can reduce the service size below 400 amps. To use the optional method, the total unit loads are multiplied by the number of units:

Lighting and Small Appliance Loads

12 x 5400 W = 64800 W

Range Load

12 x 8000 W = 96000 W

Fixed Appliances

12 x 3050 W = 36600 W

Air Conditioner

12 x 1100 W = 13200 W

Total 210600 W

From the NEC Table in Section 220-C, the total wattage for 12 to 13 dwelling units can be figured at a demand factor of 41%.

210600 x .41 = 86346 W

Minimum Feeder Size - Main to Meter Bank - Optional Method:

$$\frac{86346 \text{ W}}{230 \text{ V}} = 375 \text{ amps}$$

From the remaining 1,600 sq. ft. of the building the following loads are to be connected to the house panel:

Laundry Room:

Washer	= 1,100 W
Dryer	= 5,000 W
Exhaust Fan	= 300 W
Laundry Circuit	= 1,500 W

Office:

Air Conditioner	= 1,100 W

Halls and Lobby:

Emergency Lighting Discharge Unit	= 600 W
Exit Lights	= 300 W
Fire Alarm System	= 250 W

Mechanical Room:

3 Circulating Pumps (Sec. 430-B)	= 1,560 W (total)
3 Hot Water Heaters	= 15,000 W (total)
Fuel-Fired Boiler	= 600 W

Miscellaneous:

MATV System	= 500 W

Besides these loads, we have to figure wattage for lighting and general purpose receptacles in the non-dwelling areas. Section 220-B of the NEC allows 180 volt-amperes per outlet. A floor plan review shows that 30 outlets will give the necessary lighting and convenience receptacles inside and outside the building. Since this is primarily a resistive circuit, the volt-amperes can be said to equal the watts:

30 x 180 = 5400 W

$$\frac{5400 \text{ W}}{115 \text{ V}} = 47 \text{ amps} = \text{three 20 amp 2 wire circuits}$$

NEC demand factor rules for house loads are the same as those that apply to residences (Section 220). So, the dryer and air conditioner are taken at 100%. The exit lights, since they will be on all the time, are also taken at 100%. The remaining appliances are rated at 75%, since they are fixed and we have more than four. The laundry circuit is grouped with the general lighting load. The demand factor is 100%. See the table in 220-B.

Dryer	100% = 5000 W
Air Conditioner	100% = 1100 W
Exit Lights	100% = 300 W
General Lighting/Laundry (5400 + 1500)	100% = 6900 W

Fixed Appliances:

Washer	= 1100 W
Exhaust Fans	= 500 W
Emergency Lighting Recharger Unit	= 600 W
Fire Alarm System	= 250 W
Pumps	= 1560 W
Water Heaters	=15000 W
Boilers	= 600 W
MATV	= 500 W
Total	20110 W at 75% =15080 W

Total House Load:

	5000 W
	1100 W
	300 W
	6900 W
	28380 W

Minimum Feeder Size - Main Disconnect to House Panel

$$\frac{28,380 \text{ W}}{230 \text{ V}} = 123 \text{ amps}$$

House loads are added directly to the dwelling loads. Accordingly, the main service, its disconnect and distribution panel must have a rating of at least:

$$375 \text{ amps} + 123 \text{ amps} = 498 \text{ amps}$$

Farm Buildings Loads - Farm residence loads are figured the same as any other dwelling unit. To find the load in a farm building other than the residence, list the separate loads as follows:

1. The full-load current of the largest motor times 1.25. Where two or more motors are of equal size, apply this 125% factor to one (NEC Section 430-B). For re-rated motors, use the nameplate load.

2. The full-load currents of all other large or permanently connected items of equipment. (Usually anything larger than 1500 W or 1/2 hp or more.)

3. The convenience outlets, at 1.5 amps per outlet at 115 volts.

4. The lighting load at 1.5 amps per outlet at 115 volts unless some other factor should be applied. For example, a poultry laying house with 40 or 50 W per outlet or a building with high intensity lighting would use the actual amperage.

5. For a 115/230-volt service, add all amps taken at 230 volts. Also add one-half of all amps taken at 115 volts. Apply the demand factors found in the NEC Table in Section 220-D.

You can use Section 220-D both when finding the size of a feeder or service supplying an individual farm building and for a feeder or service to a load not in a building. The load must consist of two or more branch circuits and be 230 volts. A single item load such as an irrigation pump would have to be figured under the section on motors in this chapter.

For a 115 volt service only, add all amps at 115 volts as explained in steps 1, 2, 3, and 4 above. Total amperage should not exceed 30 and the demand factor should be 100%. If the amperage exceeds 30 amps, install 230 volt service.

Example 1 — Farm Shop 115/230 V. Service

	Amps
10 lighting outlets	
+ 6 convenience outlets	
16 outlets at 1.5 amp = 24 amp at 115 V. =	12
Heater, 3 kw; $\frac{3000 \text{ w.}}{230 \text{ V.}}$= 13 amp at 230 V. =	13
Lathe, 1 hp; 8 amp x 1.25 =	10
Saw, ¾ hp; nameplate amp rating =	6.9
Air compressor, ½ hp; nameplate amp rating =	4.9
Welder; nameplate amp rating =	35
Totals	82

Demand Factor can be figured using the Table in 220-D of the NEC:

First 60 amps at 100%	= 60 amps
Remaining 22 amps at 50%	= 11 amps
Total Load	71 amps

Example 2 — Dairy Barn 115/230 V. Service

	Amps
30 lighting outlets	
+10 convenience outlets	
40 outlets at 1.5 amp = 60 amp, 115 V. =	30
2 — 1/6 hp ventilating fans at 4.4 amps, 115 V. =	2.2
1 silo unloader, 5 hp; 28 amps x 1.25 =	35
1 feed conveyor, 5 hp =	28
Barn cleaner, 5 hp =	28

Hay elevator, 1 hp	=	8
Hay conveyor, ¾ hp	=	6.9
Milker, 1½ hp	=	10
Milk pump, ½ hp	=	4.9
Tank truck outlet, ½ hp	=	8
Milk cooler, 3 hp	=	17
Milk house heater, 2 kw	=	9
Water heater, 3 kw	=	13
Water pump, ½ hp	=	4.9
Total Amps		205

Section 220-B of the NEC lets you ignore the smaller of two dissimilar loads where it's unlikely that both loads will be used at the same time. For example, it is unlikely that the milk house heater (9 amp) and ventilating fans (2.2 amp) will be running in summer when the hay equipment is in use. So, you can subtract 11.2 amps from the load.

Be careful when applying Section 220-21. Get some advice on what equipment is used when, so you don't size the load too small.

In addition to Section 220-B, the NEC permits farm loads to be calculated using *diversity*. Diversity is a way of classifying loads which will generally not be operating at the same time. In the above example, during haying season, the hay elevator and conveyor will probably not be running while the barn cleaner is being operated. The small loads can be ignored. The barn cleaner and milker will seldom or never run at the same time.

On the other hand, the milk cooler will probably start before milking is finished. After the milker is shut off and milk house chores are being done, the water heater could come on. The milk cooler may still be running. If this were in the evening, some lights might be on. You may have to compile a list of loads that may be operating simultaneously.

Silo unloader	=	35 amp
Feed conveyor	=	25 amp
Milk cooler	=	17 amp
Water heater	=	13 amp
Lights	=	5 amp
Total loads without diversity		98 amp

We have already determined the net total load in the building in Example 2 at 194 amps. (205 less 11.2 for the milk house heater and fans equals 194 amps). Subtracting, we have 194 less 98 equals 96 amps of remaining load. Applying the Table in 220-D, 60 amps of the remainder can be figured at

50% demand factor, leaving 36 amps at a 25% demand factor.

Loads expected to operate without diversity	=	98 amp
Next 60 amp at 50% DF	=	30 amp
Remaining 36 amps at 25% DF	=	9 amp
Total computed demand load		137 amp at 230 V

With a computed demand load of 137 amps, the service entrance conductors must be rated at least at 150 amps. This provides a spare capacity of 13 amps for added loads operating without diversity. Or, 13 amps divided by 0.25 demand factor equals 52 amps of other loads operating with diversity.

If a 200 amp service were installed, spare capacity would be 200 less 137 or 63 amps. This is enough current for 2 or more 5 horsepower silo unloaders operating without diversity (2 times 28 amps equals 56 amps) plus a 5 horsepower bunk feeder at diversity (28 amps times 0.25 demand factor equals 7 amps). If all the spare capacity was used for loads operating with diversity, the spare capacity of 63 amps could serve 246 amps of added load (63 amps divided by 0.25 demand factor equals 246 amps).

The initial load which operates without diversity can't be figured at less than the first 60 amps of load or less than 125% of the full-current load of the largest motor. See NEC table in 220-D.

Selecting Conductor Size

Now you know the ampacity of the service. The next step is to select the right size conductors. Use Table 6-24.

These tables apply to both aluminum and copper conductors of various ampacities, voltages and voltage drops. The set of figures to the left of the double line gives the minimum conductor size, depending on the load, insulation, and location as required by the NEC. The figures to the right of the double line show conductor sizes based on voltage drop.

Here's an example. Let's say the dwelling requires a 100-amp, 230-volt service. A voltage drop no more than 3% is desired. The service is to be run in conduit. THW insulation is needed. In Table 6-24 (I), you see that conductors fitting these requirements must be No. 3 or larger. Go to the figures to the right. The No. 3 conductors can run up to 150 feet without exceeding 3% voltage drop.

A — Minimum Allowable Size of Conductor

Copper up to 200 Amperes, 115-120 Volts, Single Phase, Based on 2% Voltage Drop

Length of Run in Feet. Compare size shown below with size shown to left of double line. Use the larger size.

Load in Amps	Types R, T, TW	Types RH, RHW, THW	Overhead in Air* Bare & Covered Conductors	30	40	50	60	75	100	125	150	175	200	225	250	275	300	350	400	450	500	550	600	650	700
5	12	12	10	12	12	12	12	12	12	12	10	10	10	10	8	8	8	8	6	6	6	6	4	4	4
7	12	12	10	12	12	12	12	12	10	10	8	8	8	8	6	6	6	4	4	4	4	4	3
10	12	12	10	12	12	12	12	10	10	8	8	8	6	6	6	6	4	4	4	4	3	3	2	2	2
15	12	12	10	12	12	10	10	10	8	6	6	6	4	4	4	4	4	3	2	2	1	1	1	0	0
20	12	12	10	12	10	10	8	8	6	6	4	4	4	4	3	3	2	2	1	1	0	0	00	00	00
25	10	10	10	10	10	8	8	6	6	4	4	4	3	3	2	2	1	1	0	0	00	00	000	000	000
30	10	10	10	10	8	8	6	6	4	4	4	3	2	2	1	1	1	0	00	00	000	000	000	4/0	4/0
35	8	8	10	10	8	8	6	6	4	4	3	2	2	1	1	0	0	00	00	000	000	4/0	4/0	4/0	250
40	8	8	10	8	8	6	6	4	4	3	2	2	1	1	0	0	00	00	000	000	4/0	4/0	250	250	300
45	6	8	10	8	8	6	6	4	4	3	2	1	1	0	0	00	00	000	000	4/0	4/0	250	250	300	300
50	6	6	10	8	6	6	4	4	3	2	1	1	0	0	00	00	000	000	4/0	4/0	250	250	300	300	350
60	4	6	8	8	6	4	4	4	2	1	1	0	00	00	000	000	000	4/0	250	250	300	300	350	400	400
70	4	4	8	6	6	4	4	3	2	1	0	00	00	000	4/0	4/0	250	250	300	300	350	400	400	500	500
80	2	4	6	6	4	4	3	2	1	0	00	00	000	000	4/0	4/0	250	300	300	350	400	400	500	500	600
90	2	3	6	6	4	4	3	2	1	0	00	000	000	4/0	4/0	250	250	300	350	400	500	500	500	600	600
100	1	3	6	4	4	3	2	1	0	00	000	000	4/0	4/0	250	250	300	350	400	500	500	500	600	600	700
115	0	2	4	4	4	3	2	1	0	00	000	4/0	4/0	250	300	300	350	400	500	500	600	600	700	700	750
130	00	1	4	4	3	2	1	0	00	000	4/0	4/0	250	300	300	350	400	500	500	600	600	700	750	800	900
150	000	0	2	4	2	1	1	0	000	4/0	250	300	350	350	400	500	500	600	700	700	800	900	900	1M
175	4/0	00	2	3	2	1	0	00	000	4/0	250	300	350	400	400	500	500	600	700	750	800	900	1M
200	250	000	1	2	1	0	00	000	4/0	250	300	350	400	500	500	500	600	700	750	900	1M

B — Minimum Allowable Size of Conductor

Aluminum up to 200 Amperes, 115-120 Volts, Single Phase, Based on 2% Voltage Drop

Length of Run in Feet. Compare size shown below with size shown to left of double line. Use the larger size.

Load in Amps	Types R, T, TW	Types RH, RHW, THW	Overhead in Air* Bare & Covered Conductors Single Triplex	30	40	50	60	75	100	125	150	175	200	225	250	275	300	350	400	450	500	550	600	650	700
5	12	12	10	12	12	12	12	12	10	10	8	8	8	8	6	6	6	6	4	4	4	4	3	3	3
7	12	12	10	12	12	12	12	10	10	8	8	6	6	6	4	4	4	4	3	3	3	2	2	2	1
10	12	12	10	12	12	10	10	8	8	6	6	6	4	4	4	3	3	2	2	1	1	0	0	0
15	12	12	10	12	10	8	8	8	6	4	4	3	3	2	2	2	1	0	0	0	00	00	000	000
20	10	10	10	10	8	8	6	6	4	4	3	3	2	2	1	1	0	0	00	00	000	000	4/0	4/0	4/0
25	10	10	10	8	8	6	6	4	4	3	2	2	1	1	0	0	00	00	000	000	4/0	4/0	250	250	300
30	8	8	10	8	6	6	6	4	3	2	2	1	0	0	00	00	00	000	000	4/0	4/0	250	300	300	350
35	6	8	10	8	6	6	4	4	3	2	1	0	0	00	00	000	000	4/0	4/0	250	250	300	300	350	400
40	6	8	10	6	6	4	4	3	2	1	0	0	00	00	000	000	4/0	4/0	250	300	300	350	350	400	500
45	4	6	10	6	6	4	4	3	2	1	0	00	00	000	000	4/0	250	250	300	300	350	400	400	500	500
50	4	6	•8	6	4	4	3	2	1	0	00	00	000	000	4/0	4/0	250	300	300	350	400	400	500	500	600
60	2	4	6	6	4	3	3	2	0	00	00	000	4/0	4/0	250	250	300	350	350	400	500	500	600	600	700
70	2	2	6	4	4	3	2	1	0	00	000	4/0	4/0	250	300	300	350	350	500	500	600	600	700	700	750
80	1	2	6	4	3	2	1	0	00	000	4/0	4/0	250	300	300	350	350	500	500	600	600	700	750	800	900
90	0	2	4	4	3	2	1	0	00	000	4/0	250	300	300	350	400	400	500	600	600	700	750	800	900	1M
100	0	1	4	3	2	1	0	00	000	4/0	250	300	300	350	400	400	500	500	600	700	700	750	900	900	1M
115	00	0	2	3	1	1	0	00	000	4/0	300	300	350	400	500	500	600	600	700	800	900	1M
130	000	00	2	2	1	0	00	000	4/0	250	300	350	400	500	500	600	600	700	800	900	1M
150	4/0	000	1	2	0	00	00	000	250	300	350	400	500	500	600	600	700	800	900	1M
175	300	4/0	0	1	0	00	000	4/0	300	350	400	500	600	600	700	750	800	900
200	350	250	00	0	00	000	4/0	250	300	400	500	600	600	700	750	900	900

C — Minimum Allowable Size of Conductor

Copper up to 200 Amperes, 115-120 Volts, Single Phase, Based on 3% Voltage Drop

Length of Run in Feet. Compare size shown below with size shown to left of double line. Use the larger size.

Load in Amps	Types R, T, TW	Types RH, RHW, THW	Overhead in Air* Bare & Covered Conductors	30	40	50	60	75	100	125	150	175	200	225	250	275	300	350	400	450	500	550	600	650	700
5	12	12	10	12	12	12	12	12	12	12	12	12	12	10	10	10	10	8	8	8	8	8	6	6	6
7	12	12	10	12	12	12	12	12	12	12	10	10	10	8	8	8	8	6	6	6	6	4	4	4	3
10	12	12	10	12	12	12	12	12	12	10	10	8	8	8	8	6	6	6	4	4	4	4	4	4	3
15	12	12	10	12	12	12	12	10	10	8	8	6	6	6	6	4	4	4	3	3	2	2	2	1	0
20	12	12	10	12	12	12	10	10	8	8	6	6	6	4	4	4	4	3	3	2	2	2	1	1	0
25	10	10	10	12	12	10	10	8	8	6	6	6	4	4	4	3+	2	2	1	1	1	0	0	0
30	10	10	10	12	10	10	8	8	6	6	4	4	4	3	3	2	2	1	0	0	00	00	000	000
35	8	8	10	12	10	8	8	6	6	4	4	3	3	2	2	1	1	0	0	00	00	000	000	4/0
40	8	8	10	10	10	8	8	6	6	4	3	3	2	2	2	1	0	0	00	00	000	000	4/0
45	6	8	10	10	8	8	6	4	4	4	3	2	2	1	1	1	0	0	00	00	000	000	4/0	4/0
50	6	6	10	10	8	8	6	6	4	4	3	2	1	1	1	0	0	00	000	000	000	4/0	4/0	250
60	4	6	8	8	8	6	6	4	4	3	2	2	1	1	0	0	00	00	000	000	4/0	4/0	250	250	300
70	4	4	8	8	6	6	4	4	3	2	2	1	0	0	0	00	00	000	4/0	4/0	250	250	300	300	300
80	2	4	8	8	6	6	4	4	3	2	1	0	0	00	00	00	000	4/0	4/0	250	250	300	300	350	350
90	2	3	6	8	6	4	4	4	2	1	1	0	0	00	00	000	000	4/0	4/0	250	300	300	350	400	400
100	1	3	6	6	6	4	4	3	2	1	0	00	00	000	000	4/0	250	250	300	350	350	400	400	500
115	0	2	4	6	4	4	4	3	1	0	0	00	000	000	4/0	4/0	4/0	250	300	350	350	400	500	500	500
130	00	1	4	6	4	4	3	2	1	0	00	000	000	4/0	4/0	250	250	300	350	400	400	500	500	600	600
150	000	0	2	4	4	3	2	1	0	00	000	4/0	4/0	250	250	300	350	400	500	500	600	600	700	700
175	4/0	00	2	4	3	2	2	1	0	00	000	4/0	250	250	300	300	350	500	500	600	600	700	700	750
200	250	000	1	4	3	2	1	0	00	000	4/0	250	250	300	350	350	400	500	500	600	700	700	750	800	900

Conductors in overhead spans must be at least No. 10 for spans up to 50 feet and No. 8 for longer spans. See NEC, Sec. 225-6 (a). See Note 3 of NEC "Notes to Tables 310-16 through 310-19".

Conductor Sizes
Table 6-24

D — Aluminum up to 200 Amperes, 115-120 Volts, Single Phase, Based on 3% Voltage Drop

Minimum Allowable Size of Conductor. Length of Run in Feet — Compare size shown below with size shown to left of double line. Use the larger size.

Load in Amps	Types R, T, TW	Types RH, RHW, THW	Overhead in Air Bare & Covered Conductors Single Triplex	30	40	50	60	75	100	125	150	175	200	225	250	275	300	350	400	450	500	550	600	650	700
5	12	12	10	12	12	12	12	12	12	12	10	10	10	8	8	8	8	6	6	6	6	4	4	4
7	12	12	10	12	12	12	12	12	10	10	10	8	8	8	6	6	6	6	4	4	4	4	3	3
10	12	12	10	12	12	12	12	10	8	8	6	6	6	6	6	4	4	4	3	3	2	2	2	1
15	12	12	10	12	12	10	10	8	8	6	6	6	4	4	4	3	3	2	2	1	1	0	0	0
20	10	10	10	12	10	10	8	8	6	6	4	4	4	3	3	2	2	1	1	0	0	00	00	00 000
25	10	10	10	10	10	8	8	6	6	4	4	3	3	2	2	2	1	0	0	00	00	000	000	4/0
30	8	8	10	10	8	8	6	6	4	4	3	3	2	2	1	1	0	0	00	00	000	000	4/0	4/0 4/0
35	6	8	10	10	8	6	6	4	4	3	2	1	1	0	0	0	00	000	000	4/0	4/0	4/0	250	250
40	6	8	10	8	6	6	4	4	3	2	1	1	0	0	00	00	000	000	4/0	4/0	250	250	300	300
45	4	6	10	8	6	6	4	4	3	2	1	0	0	00	00	000	000	4/0	4/0	250	250	300	300	350
50	4	6	8	8	6	6	4	4	3	2	1	0	0	00	00	000	4/0	4/0	250	250	300	350	350	
60	2	4	6	6	6	4	4	3	2	1	0	0	00	00	000	000	4/0	4/0	250	300	300	350	350	400
70	2	2	6	6	4	4	4	3	1	0	0	00	000	000	4/0	4/0	4/0	250	300	350	350	400	500	500
80	1	2	6	6	4	4	3	2	1	0	00	000	000	4/0	250	250	300	350	350	400	500	500	600	
90	0	2	4	6	4	3	3	2	0	00	000	4/0	4/0	250	250	300	350	400	500	500	600	600	700	
100	0	1	4	4	4	3	2	1	0	00	000	4/0	4/0	250	250	300	300	350	400	500	500	600	600	700 700
115	00	0	2	4	3	2	1	1	00	000	000	4/0	250	300	300	350	350	400	500	600	600	700	700	750 800
130	000	00	2	4	3	2	1	0	00	000	4/0	250	300	300	350	350	400	500	600	600	700	700	800	900 900
150	4/0	000	1	3	2	1	0	00	000	4/0	250	300	300	350	400	400	500	600	600	700	750	900	900	1M
175	300	4/0	0	3	1	0	0	00	4/0	250	300	300	350	400	500	500	600	600	700	800	900	900	1M	
200	350	250	00	2	1	0	0	00	000	4/0	250	300	350	400	500	500	600	600	700	800	900	900	1M	

E — Copper up to 200 Amperes, 115-120 Volts, Single Phase, Based on 4% Voltage Drop

Minimum Allowable Size of Conductor. Length of Run in Feet — Compare size shown below with size shown to left of double line. Use the larger size.

Load in Amps	Types R, T, TW	Types RH, RHW, THW	Overhead in Air Bare & Covered Conductors	30	40	50	60	75	100	125	150	175	200	225	250	275	300	350	400	450	500	550	600	650	700
5	12	12	10	12	12	12	12	12	12	12	12	12	12	12	12	12	10	10	10	10	8	8	8	8
7	12	12	10	12	12	12	12	12	12	12	12	12	10	10	10	10	8	8	8	8	6	6	6	6
10	12	12	10	12	12	12	12	12	12	10	10	10	8	8	8	8	6	6	6	6	4	4	4	4
15	12	12	10	12	12	12	12	12	10	10	10	8	8	6	6	6	6	4	4	4	4	4	3	3
20	12	12	10	12	12	12	12	10	10	8	8	8	6	6	6	6	4	4	4	3	3	2	2	2
25	10	10	10	12	12	12	10	10	8	8	6	6	6	6	4	4	4	3	3	2	1	1	1	1
30	10	10	10	12	12	10	10	10	8	8	6	6	4	4	4	3	2	2	1	1	1	0	0	
35	8	8	10	12	12	10	10	8	8	6	4	4	4	3	3	2	2	1	1	0	0	0	00	
40	8	8	10	12	10	10	8	8	6	4	4	4	3	3	2	2	1	1	0	0	0	00	00	
45	6	8	10	12	10	10	8	8	6	6	4	4	3	2	2	1	1	0	0	00	00	00	000	
50	6	6	10	10	10	8	8	6	6	4	4	3	3	2	1	1	0	0	00	00	000	000	000	
60	4	6	8	10	8	8	6	4	4	3	2	1	1	1	0	00	00	000	000	000	4/0	4/0		
70	4	4	8	10	8	8	6	4	4	3	2	2	1	0	0	00	00	000	000	4/0	4/0	4/0	250	
80	2	4	6	8	6	6	4	4	3	2	1	1	0	0	00	00	000	000	4/0	4/0	250	300		
90	2	3	6	8	8	6	4	4	3	1	1	0	0	00	00	000	000	4/0	4/0	250	250	300	300	
100	1	3	6	8	6	6	4	4	3	2	1	1	0	0	00	00	000	000	4/0	4/0	250	250	300	300 350
115	0	2	4	8	6	6	4	4	3	2	1	0	0	00	00	000	000	4/0	4/0	250	300	300	350	350 400
130	00	1	4	6	6	4	4	3	2	1	0	0	00	00	000	000	4/0	4/0	250	300	300	350	400	400 500
150	000	0	2	6	4	4	3	1	0	0	00	000	000	4/0	4/0	4/0	250	300	350	350	400	500	500	500
175	4/0	00	2	6	4	4	3	2	1	0	00	000	000	4/0	250	250	300	300	350	400	500	500	600	600
200	250	000	1	4	4	3	2	1	0	00	000	4/0	4/0	250	300	350	350	400	500	500	600	600	700	

F — Aluminum up to 200 Amperes, 115-120 Volts, Single Phase, Based on 4% Voltage Drop

Minimum Allowable Size of Conductor. Length of Run in Feet — Compare size shown below with size shown to left of double line. Use the larger size.

Load in Amps	Types R, T, TW	Types RH, RHW, THW	Overhead in Air Bare & Covered Conductors Single Triplex	30	40	50	60	75	100	125	150	175	200	225	250	275	300	350	400	450	500	550	600	650	700
5	12	12	10	12	12	12	12	12	12	12	12	10	10	10	8	8	8	8	6	6	6	6	6	6
7	12	12	10	12	12	12	12	12	12	12	10	10	10	8	8	8	6	6	6	6	4	4	4	4
10	12	12	10	12	12	12	12	12	10	10	8	8	8	6	6	6	6	4	4	4	4	3	3	3
15	12	12	10	12	12	12	12	10	8	8	8	6	6	4	4	4	4	3	3	2	2	2	1	1
20	10	10	10	12	12	10	10	8	8	6	6	6	4	4	4	4	3	3	2	2	1	1	0	0
25	10	10	10	12	10	8	8	6	6	4	4	4	3	3	2	2	1	1	0	0	00	00	00	
30	8	8	10	12	10	8	8	6	4	4	3	3	2	2	1	1	0	0	00	00	000	000		
35	6	8	10	10	10	8	8	6	6	4	4	3	3	2	2	1	1	0	0	00	000	000	000	4/0
40	6	8	10	10	8	8	6	6	4	4	3	3	2	2	1	1	0	0	00	00	000	000	4/0	4/0
45	4	6	10	10	8	6	6	4	4	3	2	2	1	1	0	0	00	00	000	000	4/0	4/0	250	250
50	4	6	8	8	8	6	6	4	3	2	2	1	1	0	0	00	00	000	000	4/0	4/0	250	250	300
60	2	4	6	8	6	6	4	3	2	1	0	0	00	00	000	000	4/0	4/0	250	250	300	300	350	350
70	2	2	6	8	6	6	4	3	2	1	0	0	00	00	000	000	4/0	4/0	250	300	300	350	350	400
80	1	2	6	6	6	4	4	3	2	1	0	0	00	00	000	000	4/0	4/0	250	300	300	350	350	400 500
90	0	2	4	6	6	4	4	3	2	1	0	00	00	000	000	4/0	4/0	250	300	300	350	400	500	
100	0	1	4	6	4	4	3	2	1	0	00	00	000	000	4/0	4/0	250	300	300	350	400	400	500	500 600
115	00	0	2	6	4	4	3	2	1	0	00	000	000	4/0	4/0	250	300	300	350	400	500	500	600	600 600
130	000	00	2	4	4	3	2	1	0	00	000	000	4/0	250	250	300	300	350	400	500	500	600	600	700 700
150	4/0	000	1	4	3	2	2	1	00	000	000	4/0	250	250	300	300	350	400	500	500	600	600	700	750 800
175	300	4/0	0	4	3	2	1	0	00	000	4/0	250	300	300	350	400	500	500	600	600	700	750	800	900 900
200	350	250	00	3	2	1	0	00	000	4/0	250	300	300	350	400	400	500	600	600	700	750	900	900	1M

Conductors in overhead spans must be at least No. 10 for spans up to 50 feet and No. 8 for longer spans. See NEC, Sec. 225-6 (a). See Note 3 of NEC "Notes to Tables 310-16 through 310-19".

Conductor Sizes
Table 6-24 (continued)

G Minimum Allowable Size of Conductor — Copper up to 400 Amperes, 230-240 Volts, Single Phase, Based on 2% Voltage Drop

In Cable, Conduit, Earth / Overhead in Air*

Length of Run in Feet — Compare size shown below with size shown to left of double line. Use the larger size.

Load in Amps	Types R, T, TW	Types RH, RHW, THW	Bare & Covered Conductors	50	60	75	100	125	150	175	200	225	250	275	300	350	400	450	500	550	600	650	700	750	800
5	12	12	10	12	12	12	12	12	12	12	12	12	12	12	10	10	10	10	8	8	8	8	6	6
7	12	12	10	12	12	12	12	12	12	12	12	10	10	10	8	8	8	8	6	6	6	6	6	4
10	12	12	10	12	12	12	12	12	10	10	10	10	8	8	8	8	6	6	6	4	4	4	4	4
15	12	12	10	12	12	12	10	10	10	8	8	8	6	6	6	6	4	4	4	4	3	3	2	2
20	12	12	10	12	12	10	10	8	8	8	6	6	6	6	4	4	4	3	3	2	2	2	1	1
25	10	10	10	12	10	10	8	8	6	6	6	6	4	4	4	4	3	3	2	2	1	1	1	0	0
30	10	10	10	10	10	10	8	6	6	6	4	4	4	4	3	3	2	2	1	1	0	0	0	00
35	8	8	10	10	10	8	8	6	4	4	4	3	3	2	2	1	1	0	0	0	00	00	00
40	8	8	10	10	8	8	6	6	4	4	4	3	3	2	2	1	1	0	0	00	00	000	000
45	6	8	10	10	8	8	6	4	4	4	3	3	2	2	1	1	0	0	00	000	000	000	4/0
50	6	6	10	8	8	6	6	4	4	4	3	3	2	2	1	1	0	0	00	00	000	000	4/0	4/0
60	4	6	8	8	8	6	4	4	4	3	2	2	1	1	1	0	00	00	000	000	000	4/0	4/0	4/0
70	4	4	8	8	6	6	4	4	3	2	2	1	1	0	0	00	00	000	4/0	4/0	4/0	250	250	300
80	2	4	6	6	6	4	4	3	2	2	1	1	0	0	00	00	000	000	4/0	4/0	250	250	300	300	300
90	2	3	6	6	6	4	3	2	1	1	0	0	00	00	000	000	4/0	250	250	300	300	350	350
100	1	3	6	6	4	4	3	2	1	1	0	0	00	00	000	000	4/0	4/0	250	250	300	300	350	400
115	0	2	4	6	4	4	3	2	1	0	0	00	00	000	000	4/0	4/0	250	300	300	350	350	400	400	500
130	00	1	4	4	4	3	2	1	0	0	00	00	000	4/0	4/0	4/0	250	300	300	350	400	400	500	500	500
150	000	0	2	4	4	3	1	0	0	00	000	000	4/0	4/0	4/0	250	250	300	350	400	400	500	500	600	600
175	4/0	00	2	4	3	2	1	0	00	000	000	4/0	4/0	250	250	300	350	400	400	500	500	600	600	700
200	250	000	1	3	2	1	0	00	000	000	4/0	4/0	250	250	300	350	400	500	500	500	600	600	700	700	750
225	300	4/0	0	3	2	1	0	00	000	4/0	4/0	250	300	300	350	400	500	500	600	600	700	700	750	800	900
250	350	250	00	2	1	0	00	000	4/0	4/0	250	300	300	350	400	500	600	600	700	700	750	800	900	1M
275	400	300	00	2	1	0	00	000	4/0	250	250	300	350	400	500	500	600	700	700	800	900	900	1M
300	500	350	000	1	1	0	000	4/0	4/0	250	300	300	350	400	500	500	600	700	700	800	900	900	1M
325	600	400	4/0	1	0	00	000	4/0	250	300	300	350	400	500	500	600	600	700	750	900	900	1M
350	600	500	4/0	1	0	00	000	4/0	250	300	350	400	500	500	600	700	750	800	900	1M
375	700	500	250	0	0	00	4/0	250	300	300	350	400	500	500	500	600	700	800	900	1M
400	750	600	250	0	00	000	4/0	250	300	350	400	500	500	500	600	700	750	900	1M

Conductors in overhead spans must be at least No. 10 for spans up to 50 feet and No. 8 for longer spans. See NEC, Sec. 225-6 (a). See Note 3 of NEC "Notes to Tables 310-16 through 310-19"

H Minimum Allowable Size of Conductor — Aluminum up to 400 Amperes, 230-240 Volts, Single Phase, Based on 2% Voltage Drop

In Cable, Conduit, Earth / Overhead in Air*

Length of Run in Feet — Compare size shown below with size shown to left of double line. Use the larger size.

Load in Amps	Types R, T, TW	Types RH, RHW, THW	Bare & Covered Single	Bare & Covered Triplex	50	60	75	100	125	150	175	200	225	250	275	300	350	400	450	500	550	600	650	700	750	800
5	12	12	10		12	12	12	12	12	12	12	10	10	10	10	8	8	8	6	6	6	6	4	4
7	12	12	10		12	12	12	12	10	10	10	8	8	8	8	6	6	6	4	4	4	3	3	2
10	12	12	10		12	12	12	10	10	8	8	8	6	6	6	6	4	4	4	3	3	3	2	2
15	12	12	10		12	12	10	8	8	8	6	6	6	4	4	4	3	3	2	2	1	1	1	0
20	10	10	10		10	10	8	6	6	6	4	4	4	4	3	3	2	1	1	0	0	00	00
25	10	10	10		10	8	8	6	6	4	4	4	3	3	2	2	1	1	0	0	00	00	000	000
30	8	8	10		8	8	8	6	4	4	4	3	3	2	2	1	0	0	00	00	00	000	000	4/0
35	6	8	10		8	8	6	4	4	3	3	2	2	1	1	0	0	00	00	000	000	4/0	4/0	4/0
40	6	8	10		8	6	6	4	4	3	3	2	2	1	1	0	0	00	00	000	000	4/0	4/0	250	250
45	4	6	10		8	6	6	4	4	3	2	2	1	1	0	0	00	00	000	000	4/0	4/0	250	250	300
50	4	6	8		6	6	4	4	3	2	2	1	1	0	0	00	00	000	000	4/0	4/0	250	250	300	300	300
60	2	4	6	6	6	6	4	3	2	2	1	0	0	00	00	00	000	4/0	4/0	250	250	300	300	350	350	350
70	2	2(a)	6	4	6	4	4	3	2	1	0	0	00	00	00	000	4/0	4/0	250	300	300	350	400	400	500
80	1	2(a)	6	4	4	4	3	2	1	0	0	00	00	000	000	4/0	4/0	250	300	300	350	400	500	500	500
90	0	2(a)	4	2	4	4	3	2	1	0	00	00	000	000	4/0	4/0	250	300	300	350	400	400	500	500	600
100	0	1(a)	4	2	4	3	2	1	0	00	00	000	4/0	4/0	250	300	300	350	400	500	500	600	600	600
115	00	0(a)	2	1	4	3	2	1	0	00	000	000	4/0	250	300	300	350	400	500	500	600	600	700	700
130	000	00(a)	2	0	3	2	1	0	00	000	000	4/0	250	250	300	300	350	400	500	500	600	600	700	700	750	800
150	4/0	000(a)	1	00	2	2	1	00	000	4/0	250	300	300	350	400	500	500	600	600	700	700	750	900	900	900
175	300	4/0(a)	0	000	2	1	0	00	000	4/0	250	300	300	350	400	400	500	600	600	700	750	800	900	900
200	350	250	00	4/0	1	0	00	000	4/0	250	300	300	350	400	400	500	600	600	700	750	900	900	1M
225	400	300	000		1	0	000	4/0	250	300	400	400	500	500	600	700	750	900	1M	1M
250	500	350	000		0	00	000	4/0	250	300	350	400	500	500	600	700	750	900	1M
275	600	500	4/0		0	00	000	4/0	250	300	400	400	500	500	600	750	900	1M
300	700	500	250		00	0	000	250	300	350	400	500	500	600	600	700	800	900	1M
325	800	600	300		00	000	4/0	250	300	400	500	500	600	600	700	750	900	1M
350	900	700	300		00	000	4/0	300	350	400	500	600	600	700	750	800	900
375	1M	700	350		000	000	4/0	300	350	500	500	600	700	700	800	900	1M
400		900	350		000	4/0	250	300	400	500	600	600	700	750	900	900

Conductors in overhead spans must be at least No. 10 for spans up to 50 feet and No. 8 for longer spans. See NEC, Sec. 225-6 (a). See Note 3 of NEC "Notes to Tables 310-16 through 310-19".

Conductor Sizes
Table 6-24 (continued)

Table I — Copper up to 400 Amperes, 230-240 Volts, Single Phase, Based on 3% Voltage Drop

Minimum Allowable Size of Conductor

Compare size shown below with size shown to left of double line. Use the larger size.

Conductors in overhead spans must be at least No. 10 for spans up to 50 feet and No. 8 for longer spans.

Load in Amps	In Cable, Conduit, Earth — Types R, T, TW	Types RH, RHW, THW	Overhead in Air — Bare & Covered Conductors	50	60	75	100	125	150	175	200	225	250	300	350	400	450	500	550	600	650	700	800	900	1000
5	12	12	10	12	12	12	12	12	12	12	12	12	12	12	12	12	10	10	10	10	10	8	8	8	8
7	12	12	10	12	12	12	12	12	12	12	12	12	12	10	10	10	8	8	8	8	6	6	6	6
10	12	12	10	12	12	12	12	12	12	12	12	10	10	10	8	8	8	8	6	6	6	6	4	4	4
15	12	12	10	12	12	12	12	12	10	10	10	10	8	8	8	6	6	6	6	4	4	4	4	4	3
20	12	12	10	12	12	12	12	10	10	8	8	8	8	6*	6	6	4	4	4	4	4	3	3	2	2
25	10	10	10	12	12	12	10	10	8	8	8	6	6	6	4	4	4	4	3	3	2	2	1	1	1
30	10	10	10	12	12	10	10	8	8	8	6	6	6	4	4	4	3	3	2	2	2	1	1	1	0
35	8	8	10	12	12	10	8	8	8	6	6	6	6	4	4	3	3	2	2	1	1	0	0	0	
40	8	8	10	12	10	10	8	8	6	6	6	4	4	4	3	3	2	2	1	1	0	0	0	00	00
45	8	8	10	10	10	10	8	6	6	6	4	4	4	3	2	2	1	1	1	0	0	00	00	000
50	6	6	10	10	10	8	8	6	6	6	4	4	4	3	2	2	1	1	1	0	0	0	00	000	000
60	4	6	8	10	8	8	6	6	4	4	4	3	2	2	1	1	0	0	00	00	00	000	000	4/0
70	4	4	8	8	8	6	6	4	4	3	3	2	1	1	0	0	00	00	000	000	4/0	4/0	250	
80	2	4	6	8	8	6	6	4	4	3	3	2	2	1	0	0	00	00	000	000	4/0	4/0	250	250
90	2	3	6	8	8	6	4	4	4	3	2	2	1	1	0	00	00	000	000	000	4/0	4/0	250	250	300
100	1	3	6	8	6	6	4	4	4	3	2	2	1	1	0	00	00	000	000	4/0	4/0	250	250	300	350
115	0	2	4	6	6	4	4	3	3	2	1	1	0	0	00	000	000	4/0	4/0	4/0	250	250	300	350	350
130	00	1	4	6	6	4	4	3	2	1	1	0	0	00	000	000	4/0	4/0	250	250	300	300	350	400	400
150	000	0	2	6	4	4	3	2	1	0	0	00	00	000	000	4/0	4/0	250	250	300	300	350	400	500	500
175	4/0	00	2	6	4	4	2	2	1	0	0	00	00	000	4/0	250	250	300	300	350	350	400	500	500	600
200	250	000	1	4	4	3	2	1	0	0	00	000	000	4/0	250	250	300	350	350	400	400	500	500	600	700
225	300	4/0	0	4	4	3	1	0	0	00	000	000	4/0	250	300	350	350	400	500	500	500	600	600	700	700
250	350	250	00	4	3	2	1	0	00	00	000	4/0	4/0	250	300	350	400	500	500	600	600	700	700	700	800
275	400	300	00	4	3	2	1	0	00	000	000	4/0	4/0	250	300	350	400	500	500	600	600	700	800	800	900
300	500	350	000	3	2	1	0	00	000	000	4/0	250	300	350	400	500	500	600	600	700	700	750	800	900	1M
325	600	400	4/0	3	2	1	0	00	000	4/0	4/0	250	250	300	350	400	500	500	600	600	700	700	800	900	1M
350	600	500	4/0	2	2	1	0	00	000	4/0	250	250	300	350	400	500	500	600	600	700	700	750	900	1M	
375	700	500	250	2	1	0	00	000	4/0	250	250	300	350	400	500	500	600	600	700	700	750	800	900	1M	
400	750	600	250	2	1	0	00	000	4/0	250	250	300	350	400	500	500	600	600	700	700	750	800	900	1M	

Conductors in overhead spans must be at least No. 10 for spans up to 50 feet and No. 8 for longer spans. See NEC, Sec. 225-6 (a). See Note 3 of NEC "Notes to Tables 310-16 through 310-19".

Table J — Aluminum up to 400 Amperes, 230-240 Volts, Single Phase, Based on 3% Voltage Drop

Minimum Allowable Size of Conductor

Compare size shown below with size shown to left of double line. Use the larger size.

Load in Amps	In Cable, Conduit, Earth — Types R, T, TW	Types RH, RHW, THW	Overhead in Air — Single	Triplex	50	60	75	100	125	150	175	200	225	250	300	350	400	450	500	550	600	650	700	800	900	1000
5	12	12	10		12	12	12	12	12	12	12	12	12	12	10	10	10	8	8	8	8	8	6	6	6	6
7	12	12	10		12	12	12	12	12	12	12	12	10	10	10	8	8	8	6	6	6	6	6	4	4	4
10	12	12	10		12	12	12	12	12	10	10	10	8	8	6	6	6	4	4	4	4	4	3	3	3	3
15	12	12	10		12	12	12	10	10	8	8	8	8	6	6	6	4	4	4	4	3	3	3	2	2	1
20	10	10	10		12	12	10	10	8	8	6	6	6	4	4	4	3	3	2	2	2	1	1	0	0	00
25	10	10	10		12	10	10	8	8	6	6	4	4	4	3	3	2	2	1	1	1	0	0	0	00	00
30	8	8	10		10	10	8	8	6	6	6	4	4	4	3	3	2	2	1	1	0	0	0	00	00	000
35	6	8	10		10	10	8	6	6	4	4	4	3	3	2	1	1	0	0	00	00	000	000	4/0	4/0	
40	6	8	10		10	8	6	6	4	4	4	3	3	2	1	1	0	0	00	00	000	000	4/0	4/0	250	
45	4	6	10		8	8	6	6	4	4	3	3	2	2	1	0	0	00	00	00	000	000	4/0	4/0	250	250
50	4	6	8		8	8	6	6	4	4	3	2	2	1	0	0	00	00	000	000	000	4/0	4/0	250	250	250
60	2	4	6	6	8	6	6	4	4	3	3	2	2	1	0	0	00	00	000	000	4/0	4/0	4/0	250	300	300
70	2	2(a)	6	4	6	6	6	4	4	3	3	2	1	1	0	0	00	000	000	4/0	4/0	4/0	250	250	300	350
80	1	2(a)	6	4	6	6	4	4	3	3	2	1	1	0	0	00	000	000	4/0	4/0	250	250	300	300	350	400
90	0	2(a)	4	2	6	6	4	3	2	2	1	0	0	00	00	000	4/0	4/0	250	250	300	300	350	350	400	500
100	0	1(a)	4	2	6	4	4	3	2	1	0	0	00	000	000	4/0	4/0	250	250	300	300	350	350	400	500	500
115	00	0(a)	2	1	4	4	2	1	1	0	00	000	000	4/0	250	300	300	350	350	400	400	500	500	600	600
130	000	00(a)	2	0	4	4	3	2	1	0	00	00	000	000	4/0	250	300	300	350	350	400	500	500	600	600	700
150	4/0	000(a)	1	00	4	3	2	1	0	00	000	000	4/0	250	300	300	350	400	500	500	600	600	700	700	750
175	300	4/0(a)	0	000	3	3	2	0	0	00	000	4/0	4/0	250	300	300	350	400	500	500	600	600	700	700	800
200	350	250	00	4/0	3	2	1	0	00	000	4/0	4/0	250	250	300	350	400	500	500	600	600	700	700	800	900	1M
225	400	300	000		2	2	1	00	000	000	4/0	250	250	300	350	400	500	500	600	600	700	750	800	900	1M	
250	500	350	000		2	1	0	00	000	4/0	250	250	300	350	400	500	600	600	700	700	800	900	1M			
275	600	500	4/0		2	1	0	00	000	4/0	250	300	300	350	400	500	600	600	700	750	900	900	1M			
300	700	500	250		1	0	00	000	4/0	250	300	300	350	400	500	600	600	700	700	750	900	900	1M			
325	800	600	300		1	0	00	000	4/0	250	300	400	400	500	600	600	700	750	800	900	1M					
350	900	700	300		0	0	00	4/0	250	300	300	350	400	500	600	600	700	800	900	1M						
375	1M	700	350		0	00	000	4/0	250	300	350	400	500	600	700	750	900	1M								
400		900	350		0	00	000	4/0	250	300	350	400	500	500	600	700	800	900	1M							

Conductors in overhead spans must be at least No. 10 for spans up to 50 feet and No. 8 for longer spans. See NEC, Sec. 225-6 (a). See Note 3 of NEC "Notes to Tables 310-16 through 310-19".

Conductor Sizes
Table 6-24 (continued)

K — Copper up to 400 Amperes, 230-240 Volts, Single Phase, Based on 4% Voltage Drop

Minimum Allowable Size of Conductor

Compare size shown below with size shown to left of double line. Use the larger size.

Load in Amps	In Cable, Conduit, Earth — Types R, T, TW	Types RH, RHW, THW	Overhead in Air — Bare & Covered Conductors	50	60	75	100	125	150	175	200	225	250	300	350	400	450	500	550	600	650	700	800	900	1000
5	12	12	10	12	12	12	12	12	12	12	12	12	12	12	12	12	12	12	12	10	10	10	10	10	8
7	12	12	10	12	12	12	12	12	12	12	12	12	12	10	10	10	10	8	8	8	8	8	8	8
10	12	12	10	12	12	12	12	12	12	12	12	12	10	10	10	10	8	8	8	8	8	6	6	6	6
15	12	12	10	12	12	12	12	12	12	12	10	10	10	8	8	8	6	6	6	6	6	4	4	4	4
20	12	12	10	12	12	12	12	12	10	10	10	10	8	8	8	6	6	6	6	4	4	4	4	4	3
25	10	10	10	12	12	12	12	10	10	10	8	8	8	6	6	6	6	4	4	4	4	3	3	3	2
30	10	10	10	12	12	12	10	10	10	8	8	8	6	6	4	4	4	4	4	3	3	3	2	2	1
35	8	8	10	12	12	12	10	10	8	8	8	6	6	6	4	4	4	3	3	3	2	2	1	1	
40	8	8	10	12	12	10	10	8	8	8	6	6	6	4	4	4	3	3	2	2	2	1	1	0	
45	6	8	10	12	12	10	10	8	8	6	6	6	6	4	4	4	3	2	2	2	1	1	0	0	
50	6	6	10	12	10	10	8	8	6	6	6	4	4	4	3	3	2	2	1	1	1	0	0	00	
60	4	6	8	10	10	10	8	6	6	6	4	4	4	4	3	2	2	1	1	1	0	0	00	00	000
70	4	4	8	10	10	8	8	6	4	4	4	4	3	2	2	1	1	0	0	0	00	00	000		
80	2	4	6	10	8	8	6	6	4	4	4	3	2	2	1	1	0	0	00	00	000	000	4/0		
90	2	3	6	10	8	8	6	6	4	4	4	3	3	2	1	1	0	0	00	00	000	000	4/0	4/0	
100	1	3	6	8	8	6	6	4	4	4	3	3	2	1	1	0	0	00	000	000	000	4/0	4/0	250	
115	0	2	4	8	8	6	4	4	3	3	2	2	1	0	0	00	00	000	000	4/0	4/0	4/0	250	300	
130	00	1	4	8	6	6	4	4	3	3	2	2	1	0	0	00	000	000	4/0	4/0	250	250	300	300	
150	000	0	2	6	6	6	4	3	3	2	1	1	0	00	00	000	000	4/0	4/0	250	250	300	350	350	
175	4/0	00	2	6	6	4	4	3	2	1	1	0	0	00	000	000	4/0	4/0	250	250	300	300	350	400	
200	250	000	1	6	4	4	3	2	1	1	0	0	00	000	000	4/0	4/0	250	250	300	300	350	400	500	
225	300	4/0	0	6	4	4	3	2	1	0	0	00	000	000	4/0	4/0	250	300	300	350	350	350	400	600	
250	350	250	00	4	4	3	2	1	0	0	00	000	4/0	4/0	250	300	300	350	350	400	400	500	600		
275	400	300	00	4	4	3	2	1	0	00	000	000	4/0	250	250	300	300	350	400	500	500	500	600	700	
300	500	350	000	4	4	3	1	0	0	000	000	4/0	4/0	250	300	300	350	350	400	500	500	600	700	700	
325	600	400	4/0	4	3	2	1	0	00	00	000	000	4/0	250	300	300	350	400	500	500	500	600	600	700	750
350	600	500	4/0	4	3	2	1	0	00	000	000	4/0	4/0	250	300	350	400	400	500	500	600	600	700	750	800
375	700	500	250	3	3	2	0	00	000	000	4/0	4/0	250	300	300	350	400	500	500	600	600	700	700	800	900
400	750	600	250	3	2	1	0	00	000	000	4/0	4/0	250	300	350	400	500	500	500	600	600	700	750	900	1M

Conductors in overhead spans must be at least No. 10 for spans up to 50 feet and No. 8 for longer spans. See NEC, Sec. 225-6 (a).
See Note 3 of NEC "Notes to Tables 310-16 through 310-19".

L — Aluminum up to 400 Amperes, 230-240 Volts, Single Phase, Based on 4% Voltage Drop

Minimum Allowable Size of Conductor

Compare size shown below with size shown to left of double line. Use the larger size.

Load in Amps	In Cable Conduit, Earth — Types R, T, TW	Types RH, RHW, THW	Overhead in Air — Single	Triplex	50	60	75	100	125	150	175	200	225	250	300	350	400	450	500	550	600	650	700	800	900	1000
5	12	12	10		12	12	12	12	12	12	12	12	12	12	12	12	10	10	10	10	8	8	8	8	8	6
7	12	12	10		12	12	12	12	12	12	12	12	12	10	10	10	8	8	8	8	6	6	6	6	6	4
10	12	12	10		12	12	12	12	12	12	10	10	10	8	8	8	6	6	6	6	6	4	4	4	4	
15	12	12	10		12	12	12	10	10	10	8	8	8	8	6	6	6	4	4	4	4	3	3	2		
20	10	10	10		12	12	12	10	8	8	8	6	6	6	6	4	4	4	3	3	2	2	1			
25	10	10	10		12	12	10	10	8	8	6	6	6	6	4	4	4	3	3	2	2	2	1	1	0	
30	8	8	10		12	12	10	8	8	8	6	6	6	4	4	3	3	2	2	1	1	0	0	00		
35	6	8	10		12	12	10	8	8	6	6	6	4	4	4	3	3	2	1	1	1	0	0	00	00	
40	6	8	10		10	10	8	8	6	6	4	4	4	3	3	2	2	1	1	0	0	0	00	00	000	
45	4	6	10		10	10	8	8	5	6	4	4	4	3	2	2	1	1	0	0	00	00	00	000	000	4/0
50	4	6	8		10	8	8	6	6	4	4	4	3	2	2	1	1	0	0	00	00	000	000	4/0		
60	2	4	6	6	8	8	8	6	4	4	4	3	3	2	1	0	0	00	00	000	000	4/0	4/0	250		
70	2	2(a)	6	4	8	8	6	6	4	4	3	3	2	2	1	0	0	00	00	000	000	000	4/0	4/0	250	300
80	1	2(a)	6	4	8	6	6	4	4	3	3	2	2	1	0	0	00	00	000	4/0	4/0	4/0	250	300	300	
90	0	2(a)	4	2	8	6	6	4	4	3	2	2	1	1	0	00	00	000	000	4/0	4/0	250	250	300	300	350
100	0	1(a)	4	2	6	6	4	4	3	2	1	1	0	00	00	000	000	4/0	4/0	250	250	300	300	350		
115	00	0(a)	2	1	6	6	4	4	3	2	1	1	0	0	00	000	000	4/0	4/0	250	300	300	300	350	400	500
130	000	00(a)	2	0	6	4	4	3	2	1	1	0	00	00	000	000	4/0	250	250	300	300	350	350	400	500	500
150	4/0	000(a)	1	00	4	4	4	2	1	1	0	00	00	000	000	4/0	250	250	300	300	350	400	400	500	500	600
175	300	4/0(a)	0	000	4	4	3	2	1	0	00	00	000	000	4/0	250	300	300	350	400	500	500	600	600	600	700
200	350	250	00	4/0	4	3	2	1	0	00	00	000	000	4/0	250	300	300	350	400	400	500	500	600	600	700	750
225	400	300	000		4	3	2	1	0	00	000	000	4/0	4/0	250	300	350	400	500	500	500	600	600	700	750	900
250	500	350	000		3	2	1	0	00	000	000	4/0	4/0	250	300	300	400	500	500	500	600	600	700	750	900	1M
275	600	500	4/0		3	2	1	0	00	000	4/0	4/0	250	250	300	400	400	500	500	600	600	700	750	900	1M	
300	700	500	250		2	2	1	00	000	000	4/0	250	300	350	400	500	500	500	600	600	700	700	800	900	1M	
325	800	600	300		2	1	0	00	000	4/0	4/0	250	300	300	400	500	500	600	600	700	750	800	900	1M		
350	900	700	300		2	1	0	00	000	4/0	250	300	300	350	400	500	600	600	700	750	800	900	900			
375	1M	700	350		1	1	0	000	4/0	4/0	250	300	350	350	500	500	600	700	700	800	900	900	1M			
400		900	350		1	0	00	000	4/0	250	300	300	350	400	500	600	600	700	750	900	900	1M				

Conductors in overhead spans must be at least No. 10 for spans up to 50 feet and No. 8 for longer spans. See NEC, Sec. 225-6 (a).
See Note 3 of NEC "Notes to Tables 310-16 through 310-19".

Conductor Sizes
Table 6-24 (continued)

M — Minimum Allowable Size of Conductor

Copper up to 400 Amperes, 480 Volts, Single Phase, Based on 2% Voltage Drop

Length of Run in Feet. Compare size shown below with size shown to left of double line. Use the larger size.

Load in Amps	Types R, T, TW	Types RH, RHW, THW	Bare & Covered Conductors (Overhead in Air)	100	150	200	250	300	350	400	450	500	550	600	650	700	750	800	900	1000	1200	1400	1600	1800	2000
5	12	12	10	12	12	12	12	12	12	12	12	12	12	10	10	10	10	10	10	8	8	8	6	6	6
7	12	12	10	12	12	12	12	12	12	12	10	10	10	10	10	8	8	8	8	6	6	6	4	4	4
10	12	12	10	12	12	12	12	10	10	10	10	8	8	8	8	8	6	6	6	6	4	4	4	4	3
15	12	12	10	12	12	10	10	10	8	8	8	6	6	6	6	6	6	4	4	4	4	3	2	2	1
20	12	12	10	12	10	10	8	8	8	6	6	6	6	4	4	4	4	4	3	2	2	1	1	0	
25	10	10	10	12	10	8	8	6	6	6	6	4	4	4	4	3	3	3	2	1	1	0	0	00	
30	10	10	10	10	10	8	6	6	6	4	4	4	4	3	3	3	2	2	1	1	0	00	00	000	
35	8	8	10	10	8	8	6	6	4	4	4	4	3	3	3	2	2	2	1	1	0	00	00	000	
40	8	8	10	10	8	6	6	4	4	4	4	3	3	2	2	2	1	1	1	0	00	00	000	000	4/0
45	6	8	10	10	8	6	6	4	4	4	3	2	2	2	1	1	1	0	00	00	000	000	4/0	4/0	
50	6	6	10	8	6	6	4	4	4	3	3	2	2	1	1	1	0	0	0	00	000	000	4/0	4/0	250
60	4	6	8	8	6	4	4	4	3	2	2	1	1	1	0	0	0	00	00	000	000	4/0	250	250	300
70	4	6	8	8	6	4	4	3	2	2	1	1	0	0	0	00	00	00	000	000	4/0	250	300	300	350
80	2	4	6	6	4	4	3	2	2	1	1	0	0	00	00	00	000	000	4/0	250	300	300	350	400	
90	2	3	6	6	4	4	3	2	1	1	0	0	00	00	000	000	000	4/0	4/0	250	300	350	400	500	
100	1	3	6	6	4	3	2	1	1	0	0	00	000	000	000	4/0	4/0	250	300	350	400	500	500		
115	0	2	4	6	4	3	2	1	0	0	00	00	000	000	4/0	4/0	4/0	4/0	250	300	350	400	500	500	600
130	00	1	4	4	3	2	1	0	0	00	00	000	000	4/0	4/0	4/0	250	250	300	300	400	500	500	600	
150	000	0	2	4	3	1	0	0	00	00	000	000	4/0	4/0	250	250	300	300	350	350	400	500	600	700	700
175	4/0	00	2	4	2	1	0	00	000	000	4/0	4/0	250	250	300	300	350	400	400	500	600	700	750	800	
200	250	000	1	3	1	0	00	000	000	4/0	4/0	250	250	300	300	350	350	400	500	500	600	700	750	900	1M
225	300	4/0	0	3	1	0	4/0	4/0	250	300	300	350	350	400	400	400	500	500	600	700	750	900	1M		
250	350	250	00	2	0	00	000	4/0	4/0	250	300	300	350	350	400	400	500	500	600	600	700	800	1M		
275	400	300	00	2	0	00	000	4/0	250	250	300	350	350	400	500	500	500	500	600	700	800	900			
300	500	350	000	1	0	0	000	4/0	250	300	350	350	400	500	500	500	600	600	700	700	900	1M			
325	600	400	4/0	1	00	000	4/0	250	300	300	350	400	500	500	600	600	600	600	700	750	900				
350	600	500	4/0	1	00	000	4/0	250	300	350	400	400	500	500	600	600	600	700	750	800	1M				
375	700	500	250	0	00	4/0	250	300	300	350	400	500	500	500	600	600	700	700	800	900					
400	750	600	250	0	000	4/0	250	300	350	400	400	500	500	500	600	600	700	700	750	900	1M				

Conductors in overhead spans must be at least No. 10 for spans up to 50 feet and No. 8 for longer spans. See NEC, Sec. 225-6 (a). See Note 3 of NEC "Notes to Tables 310-16 through 310-19".

N — Minimum Allowable Size of Conductor

Aluminum up to 400 Amperes, 480 Volts, Single Phase, Based on 2% Voltage Drop

Length of Run in Feet. Compare size shown below with size shown to left of double line. Use the larger size.

Load in Amps	Types R, T, TW	Types RH, RHW, THW	Bare & Covered (Single)	Bare & Covered (Triplex)	100	150	200	250	300	350	400	450	500	550	600	650	700	750	800	900	1000	1200	1400	1600	1800	2000
5	12	12	10		12	12	12	12	12	12	10	10	10	10	8	8	8	8	8	8	6	6	6	4	4	4
7	12	12	10		12	12	12	12	10	10	10	8	8	8	8	8	6	6	6	6	6	4	4	4	3	3
10	12	12	10		12	12	10	10	8	8	8	6	6	6	6	4	4	4	4	4	3	3	2	2	1	
15	12	12	10		12	10	8	8	8	6	6	6	4	4	4	4	4	3	3	2	2	1	0	0	00	
20	10	10	10		10	8	8	6	6	6	4	4	4	4	3	3	3	2	2	2	1	0	0	00	00	000
25	10	10	10		10	8	6	4	4	4	4	4	3	3	2	2	1	1	1	1	0	00	00	000	000	4/0
30	8	8	10		8	6	6	4	4	3	3	2	2	2	1	1	1	0	00	00	000	4/0	4/0	250		
35	6	8	10		8	6	4	4	3	3	2	2	1	1	1	0	0	00	00	000	4/0	4/0	250	300		
40	6	8	10		8	6	4	4	3	3	2	2	1	1	0	0	0	00	00	000	000	4/0	4/0	250	300	300
45	4	6	10		8	6	4	4	3	2	2	1	1	0	0	00	00	00	000	000	4/0	250	300	300	350	
50	4	6	8		6	4	4	3	2	1	1	0	0	00	00	000	000	000	4/0	250	300	300	350	400		
60	2	4	6	6	6	4	3	2	2	1	0	0	00	00	00	000	000	000	4/0	250	300	350	350	400	500	
70	2	2(a)	6	4	6	4	3	2	1	0	0	00	00	000	000	000	4/0	4/0	4/0	250	300	350	400	500	500	600
80	1	2(a)	6	4	4	3	2	1	0	00	00	000	000	4/0	4/0	4/0	4/0	250	250	300	300	350	350	500	500	600
90	0	2(a)	4	2	4	3	2	1	0	00	00	000	000	4/0	4/0	250	250	250	300	300	350	400	500	600	600	700
100	0	1(a)	4	2	4	2	1	0	00	00	000	000	4/0	4/0	250	300	300	300	350	400	500	600	600	700	750	
115	00	0(a)	2	1	4	2	1	0	00	000	000	4/0	4/0	250	300	300	300	350	400	500	600	600	700	700	900	
130	000	00(a)	2	0	3	1	0	00	000	000	4/0	250	250	300	300	350	350	400	400	500	500	600	700	800	900	1M
150	4/0	000(a)	1	00	2	1	00	000	000	4/0	250	250	300	300	350	400	400	500	500	500	600	700	800	900		
175	300	4/0(a)	0	000	2	0	00	000	4/0	250	300	300	350	350	400	400	500	500	600	600	600	700	800	900		
200	350	250	00	4/0	1	00	000	4/0	250	300	350	350	400	400	500	500	600	600	600	700	750	900				
225	400	300	000		1	00	000	4/0	250	300	350	400	500	500	500	600	600	700	700	750	900	1M				
250	500	350	000		0	000	4/0	250	300	350	400	500	500	500	600	600	700	700	750	750	1M					
275	600	500	4/0		0	000	4/0	250	300	400	500	500	500	600	600	600	700	750	800	900	1M					
300	700	500	250		00	000	250	300	350	400	500	500	600	600	700	750	800	900	900	1M						
325	800	600	300		00	4/0	250	300	500	500	600	600	700	750	800	900	900	1M								
350	900	700	300		00	4/0	300	350	400	500	500	600	700	750	800	900	900	1M								
375	1M	700	350		000	4/0	300	350	500	500	600	700	700	800	900	900	1M									
400		900	350		000	250	300	400	500	600	600	700	750	900	900	1M										

Conductors in overhead spans must be at least No. 10 for spans up to 50 feet and No. 8 for longer spans. See NEC, Sec. 225-6 (a). See Note 3 of NEC "Notes to Tables 310-16 through 310-19".

Conductor Sizes
Table 6-24 (continued)

O — Minimum Allowable Size of Conductor

Copper up to 400 Amperes, 480 Volts, Single Phase, Based on 3% Voltage Drop

Length of Run in Feet — Compare size shown below with size shown to left of double line. Use the larger size.

Load in Amps	Types R, T, TW	Types RH, RHW, THW	Overhead in Air* Bare & Covered Conductors	100	150	200	250	300	350	400	450	500	550	600	650	700	750	800	900	1000	1200	1400	1600	1800	2000
5	12	12	10	12	12	12	12	12	12	12	12	12	12	12	12	12	12	12	10	10	10	8	8	8	8
7	12	12	10	12	12	12	12	12	12	12	12	12	12	12	10	10	10	10	10	8	8	8	6	6	6
10	12	12	10	12	12	12	12	12	12	10	10	10	10	10	8	8	8	8	8	6	6	6	6	4	4
15	12	12	10	12	12	12	12	10	10	10	10	8	8	8	8	6	6	6	6	4	4	3	3	2	2
20	12	12	10	12	12	12	12	10	10	8	8	8	8	6	6	6	6	6	6	4	4	3	3	2	2
25	10	10	10	12	12	10	10	8	8	8	6	6	6	6	6	4	4	4	4	3	2	2	1	1	1
30	10	10	10	12	10	10	8	8	8	6	6	6	6	4	4	4	4	4	3	2	2	1	1	0	0
35	8	8	10	12	10	8	8	6	6	6	6	4	4	4	4	4	3	3	2	1	1	0	0	0	0
40	8	8	10	12	10	8	8	6	6	6	4	4	4	4	4	3	3	3	2	2	1	0	0	00	00
45	6	8	10	10	10	8	6	6	6	4	4	4	4	4	3	3	3	2	2	1	1	0	00	00	00
50	6	6	10	10	8	8	6	6	6	4	4	4	4	3	3	3	2	2	2	1	1	0	00	000	000
60	4	6	8	10	8	6	6	4	4	4	4	3	3	2	2	2	1	1	1	0	00	00	000	000	4/0
70	4	4	8	8	8	6	6	4	4	3	3	2	2	1	1	1	0	0	0	00	000	4/0	4/0	250	250
80	2	4	6	8	6	6	4	4	3	3	2	2	2	1	1	0	0	0	00	000	000	4/0	4/0	250	250
90	2	3	6	8	6	4	4	4	3	2	2	1	1	1	0	0	0	00	00	000	4/0	4/0	250	250	300
100	1	3	6	8	6	4	4	3	2	2	1	1	1	0	0	0	00	00	000	000	4/0	250	250	300	350
115	0	2	4	6	6	4	3	3	2	1	1	0	0	0	00	00	000	000	4/0	4/0	250	300	350	350	350
130	00	1	4	6	4	4	3	2	1	1	0	0	00	00	00	000	000	000	4/0	4/0	250	300	350	400	400
150	000	0	2	6	4	4	3	2	1	1	0	00	00	00	000	000	4/0	4/0	4/0	250	300	350	400	500	500
175	4/0	00	2	6	4	2	2	1	0	0	00	000	000	4/0	4/0	4/0	250	250	300	350	400	500	500	600
200	250	000	1	4	3	2	1	0	0	00	000	000	000	4/0	4/0	250	250	250	300	350	400	500	500	600	700
225	300	4/0	0	4	3	1	0	00	000	000	4/0	4/0	4/0	4/0	250	250	300	300	350	350	400	500	500	600	700
250	350	250	00	4	2	1	0	00	00	000	4/0	4/0	4/0	250	250	300	300	350	350	400	500	500	600	700	700
275	400	300	00	4	2	1	0	00	000	000	4/0	4/0	250	250	300	300	350	350	400	500	500	600	700	800
300	500	350	000	3	1	0	00	000	000	4/0	250	250	300	300	350	350	400	500	500	600	700	750	900	1M
325	600	400	4/0	3	1	0	00	000	4/0	4/0	250	250	300	300	350	350	400	400	500	600	700	800	900	1M
350	600	500	4/0	2	1	0	00	000	4/0	250	250	300	300	350	350	400	400	500	500	600	700	750	900	1M
375	700	500	250	2	0	00	000	4/0	4/0	250	250	300	350	350	400	400	500	500	600	700	800	1M
400	750	600	250	2	0	00	000	4/0	250	250	300	350	350	400	400	500	500	600	700	750	900	1M

Conductors in overhead spans must be at least No. 10 for spans up to 50 feet and No. 8 for longer spans. See NEC, Sec. 225-6 (a). See Note 3 of NEC "Notes to Tables 310-16 through 310-19"

P — Minimum Allowable Size of Conductor

Aluminum up to 400 Amperes, 480 Volts, Single Phase, Based on 3% Voltage Drop

Length of Run in Feet — Compare size shown below with size shown to left of double line. Use the larger size.

Load in Amps	Types R, T, TW	Types RH, RHW, THW	Overhead in Air* Single	Overhead in Air* Triplex	100	150	200	250	300	350	400	450	500	550	600	650	700	750	800	900	1000	1200	1400	1600	1800	2000
5	12	12	10		12	12	12	12	12	12	12	12	12	12	10	10	10	10	10	8	8	8	6	6	6	6
7	12	12	10		12	12	12	12	12	12	10	10	10	10	10	8	8	8	8	8	6	6	6	4	4	4
10	12	12	10		12	12	12	12	10	10	10	8	8	8	8	6	6	6	6	6	4	4	4	3	3	3
15	12	12	10		12	12	10	10	8	8	8	8	6	6	6	6	4	4	4	4	3	3	2	2	1	
20	10	10	10		12	10	10	8	8	6	6	6	6	4	4	4	4	4	3	3	2	1	1	0	0	
25	10	10	10		12	10	8	8	6	6	4	4	4	4	4	3	3	2	2	1	1	0	0	00	00	000
30	8	8	10		10	8	8	6	6	6	4	4	4	3	3	3	2	2	2	1	0	0	0	00	00	000
35	6	8	10		10	8	6	6	4	4	4	3	3	2	2	2	1	1	1	0	0	00	000	000	000	4/0
40	6	8	10		10	8	6	6	4	4	3	3	2	2	2	1	1	1	0	0	00	000	000	4/0	4/0	
45	4	6	10		8	8	6	4	4	4	3	3	2	2	1	1	1	0	0	00	00	000	4/0	4/0	250	
50	4	6	8		8	6	6	4	4	3	3	2	2	1	1	1	0	0	00	00	000	4/0	4/0	250	250	
60	2	4	6	6	8	6	4	4	3	3	2	2	1	1	0	0	0	00	00	000	4/0	4/0	250	300	300	
70	2	2(a)	6	4	6	6	4	3	3	2	1	1	0	0	0	00	00	00	000	000	4/0	4/0	250	300	350	350
80	1	2(a)	6	4	6	4	4	3	2	1	1	0	0	00	00	000	000	000	4/0	4/0	250	300	350	350	400	
90	0	2(a)	4	2	6	4	3	2	2	1	0	0	00	00	00	000	000	000	4/0	4/0	250	300	350	350	400	500
100	0	1(a)	4	2	6	4	3	2	1	0	0	00	00	00	000	000	4/0	4/0	4/0	250	250	300	350	400	500	500
115	00	0(a)	2	1	4	4	2	1	1	0	00	00	000	000	000	4/0	4/0	250	300	300	350	400	500	500	600	600
130	000	00(a)	2	0	4	3	2	1	0	00	000	000	4/0	4/0	4/0	250	250	300	300	350	400	500	600	600	700	
150	4/0	000(a)	1	00	4	2	1	0	00	00	000	000	4/0	4/0	250	250	300	300	350	400	500	600	600	700	750	
175	300	4/0(a)	0	000	3	2	0	0	000	000	4/0	250	250	300	300	350	350	400	500	500	600	700	800	900		
200	350	250	00	4/0	3	1	0	00	000	4/0	4/0	250	250	300	300	350	350	400	400	500	500	600	700	800	900	1M
225	400	300	000		2	1	00	000	000	4/0	250	250	300	300	350	400	400	500	500	500	600	700	800	900	1M	
250	500	350	000		2	0	00	000	4/0	250	300	350	350	400	400	500	500	600	600	700	750	900	1M			
275	600	500	4/0		2	0	00	000	4/0	250	300	300	350	400	400	500	500	600	600	700	900	1M				
300	700	500	250		1	00	000	4/0	250	300	300	350	400	400	500	500	600	600	700	750	900					
325	800	600	300		1	00	000	4/0	250	300	400	400	500	500	600	600	600	700	750	800	1M					
350	900	700	300		0	00	4/0	250	300	300	350	400	500	500	600	600	700	700	800	900						
375	1M	700	350		0	000	4/0	250	300	350	400	500	500	600	600	700	700	750	900	1M						
400		900	350		0	000	4/0	250	300	350	400	500	500	600	600	700	700	750	800	900	1M					

Conductors in overhead spans must be at least No. 10 for spans up to 50 feet and No. 8 for longer spans. See NEC, Sec. 225-6 (a). See Note 3 of NEC "Notes to Tables 310-16 through 310-19".

Conductor Sizes
Table 6-24 (continued)

Q — Minimum Allowable Size of Conductor — Copper up to 400 Amperes, 480 Volts, Single Phase, Based on 4% Voltage Drop

Load in Amps	In Cable, Conduit, Earth — Types R, T, TW	In Cable, Conduit, Earth — Types RH, RHW, THW	Overhead in Air° — Bare & Covered Conductors	100	150	200	250	300	350	400	450	500	550	600	650	700	750	800	900	1000	1200	1400	1600	1800	2000
5	12	12	10	12	12	12	12	12	12	12	12	12	12	12	12	12	12	12	12	10	10	10	10	8
7	12	12	10	12	12	12	12	12	12	12	12	12	12	12	12	12	12	12	10	10	10	8	8	8
10	12	12	10	12	12	12	12	12	12	12	12	12	10	10	10	10	10	10	10	8	8	6	6	6
15	12	12	10	12	12	12	12	12	12	10	10	10	10	10	8	8	8	8	6	6	6	4	4	4
20	12	12	10	12	12	12	12	10	10	10	10	8	8	8	8	6	6	6	6	4	4	4	4	3
25	10	10	10	12	12	12	10	10	10	8	8	8	8	6	6	6	6	4	4	4	4	3	3	2
30	10	10	10	12	12	10	10	10	8	8	8	6	6	6	6	6	4	4	4	3	2	2	2	1
35	8	8	10	12	12	10	10	8	8	8	6	6	6	6	4	4	4	4	3	2	2	1	1	1
40	8	8	10	12	10	10	8	8	8	6	6	6	4	4	4	4	4	3	2	2	1	1	1	0
45	6	8	10	12	10	10	8	6	6	6	6	4	4	4	4	4	3	3	2	1	1	0	0	0
50	6	6	10	12	10	8	8	6	6	6	6	4	4	4	4	4	3	3	2	1	1	0	0	00
60	4	6	8	10	10	8	6	6	6	4	4	4	4	3	3	3	2	2	1	1	0	00	00	000
70	4	6	8	10	8	8	6	6	4	4	4	3	3	3	2	2	2	1	1	0	00	00	000	000
80	2	4	6	10	8	6	6	4	4	4	4	3	3	2	2	2	1	1	0	00	00	000	000	4/0
90	2	3	6	10	8	6	6	4	4	4	3	3	2	2	2	1	1	1	0	0	00	000	000	4/0
100	1	3	6	8	6	6	4	4	3	3	3	2	2	1	1	1	0	0	00	000	000	4/0	4/0	250
115	0	2	4	8	6	6	4	4	3	3	2	2	1	1	0	0	0	00	00	000	4/0	4/0	4/0	300
130	00	1	4	8	6	4	4	3	3	2	2	1	1	0	0	0	00	00	000	000	4/0	4/0	250	300
150	000	0	2	6	6	4	3	2	1	1	1	0	0	0	00	00	00	000	000	4/0	4/0	250	300	300
175	4/0	00	2	6	4	4	3	2	1	1	0	0	00	00	00	000	000	4/0	4/0	250	300	350	350	350
200	250	000	1	6	4	3	2	1	1	0	0	00	00	000	000	000	4/0	4/0	4/0	250	300	350	400	500
225	300	4/0	0	6	4	3	2	1	0	0	00	00	000	000	000	4/0	4/0	4/0	250	300	350	400	500	500	600
250	350	250	00	4	3	2	1	0	0	00	00	000	000	4/0	4/0	4/0	250	300	350	400	500	500	600
275	400	300	00	4	3	2	1	0	00	00	000	000	4/0	4/0	4/0	250	250	300	300	350	400	500	600	600
300	500	350	000	4	3	1	0	0	00	000	000	4/0	4/0	4/0	250	250	300	300	350	350	500	500	600	700	700
325	600	400	4/0	4	2	1	0	00	00	000	4/0	4/0	250	250	300	300	350	350	400	500	600	600	700	750
350	600	500	4/0	4	2	1	0	00	000	000	4/0	4/0	250	250	300	300	350	400	400	500	600	600	700	750
375	700	500	250	3	2	0	00	000	4/0	4/0	250	250	300	300	300	350	350	400	500	600	600	700	800	900
400	750	600	250	3	1	0	00	000	000	4/0	4/0	250	250	300	300	300	350	350	400	500	600	600	700	750	900	1M

Conductors in overhead spans must be at least No. 10 for spans up to 50 feet and No. 8 for longer spans. See NEC, Sec. 225-6 (a). See Note 3 of NEC "Notes to Tables 310-16 through 310-19".

R — Minimum Allowable Size of Conductor — Aluminum up to 400 Amperes, 480 Volts, Single Phase, Based on 4% Voltage Drop

Load in Amps	In Cable Conduit, Earth — Types R, T, TW	In Cable Conduit, Earth — Types RH, RHW, THW	Overhead in Air° — Single	Overhead in Air° — Triplex	100	150	200	250	300	350	400	450	500	550	600	650	700	750	800	900	1000	1200	1400	1600	1800	2000
5	12	12	10		12	12	12	12	12	12	12	12	12	12	12	12	12	10	10	10	10	8	8	8	8	6
7	12	12	10		12	12	12	12	12	12	12	12	12	12	10	10	10	10	10	10	8	8	8	6	6	6
10	12	12	10		12	12	12	12	12	12	10	10	10	10	8	8	8	8	8	8	6	6	6	4	4	4
15	12	12	10		12	12	12	10	10	10	8	8	8	8	6	6	6	6	6	6	4	4	4	3	3	2
20	10	10	10		12	12	10	10	8	8	8	8	6	6	6	6	4	4	4	4	3	3	3	2	2	1
25	10	10	10		12	10	10	8	8	8	6	6	6	6	4	4	4	4	4	3	2	2	1	1	0
30	8	8	10		12	10	8	8	6	6	6	6	4	4	4	4	4	3	3	2	2	1	0	0	00
35	6	8	10		12	10	8	8	6	6	4	4	4	4	3	3	3	2	2	1	0	0	00	00
40	6	8	10		10	8	8	6	6	4	4	4	3	3	2	2	2	1	0	0	00	00	000
45	4	6	10		10	8	8	6	6	4	4	4	3	3	2	2	2	1	1	0	00	00	000
50	4	6	8		10	8	6	6	4	4	4	3	3	2	2	2	1	1	1	0	00	00	000	000	4/0
60	2	4	6	6	8	6	4	4	3	2	2	2	1	1	1	0	0	00	00	000	4/0	4/0	250
70	2	2(a)	6	4	8	6	6	4	3	3	2	1	1	0	0	0	00	00	000	000	4/0	250	300
80	1	2(a)	6	4	8	6	4	4	3	3	2	1	1	0	0	0	00	00	000	000	4/0	250	300
90	0	2(a)	4	2	8	6	4	4	3	2	2	1	1	0	0	00	00	00	000	000	4/0	250	300	350
100	0	1(a)	4	2	6	4	4	3	2	2	1	1	0	0	00	00	000	000	000	4/0	250	300	350	400
115	00	0(a)	2	1	6	4	4	3	2	1	1	0	0	00	00	000	000	000	000	4/0	4/0	300	300	350	400	500
130	000	00(a)	2	0	6	4	3	2	1	1	0	00	00	000	000	000	4/0	4/0	250	250	300	350	400	500	500
150	4/0	000(a)	1	00	4	4	2	1	1	0	00	00	000	000	4/0	4/0	4/0	250	250	300	350	400	500	500	600
175	300	4/0(a)	0	000	4	3	2	1	0	00	00	000	000	4/0	4/0	4/0	250	250	300	300	350	400	500	600	600	700
200	350	250	00	4/0	4	2	1	0	00	00	000	000	4/0	4/0	250	250	300	300	350	350	400	500	600	700	750
225	400	300	000		4	2	1	0	00	000	000	4/0	4/0	250	250	300	300	350	350	400	500	600	700	750	900
250	500	350	000		3	1	0	00	000	000	4/0	4/0	250	250	300	300	350	400	400	500	600	700	750	900	1M
275	600	500	4/0		3	1	0	00	000	000	4/0	4/0	250	250	300	300	400	400	400	500	600	750	900	1M
300	700	500	250		2	1	00	000	000	4/0	250	250	300	300	350	400	400	500	500	500	600	700	800	900	1M
325	800	600	300		2	0	00	000	4/0	4/0	250	300	300	350	400	400	500	500	500	600	600	750	900	1M
350	900	700	300		2	0	00	000	4/0	250	300	300	350	400	400	500	500	500	600	700	900	900
375	1M	700	350		1	0	000	4/0	4/0	250	300	350	350	400	500	500	600	600	700	700	900	1M
400		900	350		1	00	000	4/0	250	300	300	350	400	400	500	500	600	600	700	750	900

Conductors in overhead spans must be at least No. 10 for spans up to 50 feet and No. 8 for longer spans. See NEC, Sec. 225-6 (a). See Note 3 of NEC "Notes to Tables 310-16 through 310-19".

Conductor Sizes
Table 6-24 (continued)

Note that for shorter distances, smaller size conductors meet the voltage drop requirements. However, they would not fill the NEC minimum size. So, they can't be used.

Also keep in mind that conductors in overhead spans must be at least No. 10 wire up to 50 feet. Longer spans require at least No. 8 wire. They must be kept at minimum distances above roofs, buildings and roadways. See NEC Section 230.

The general rule is that only one service is permitted for any one building or structure. Exceptions are granted for very large buildings, emergency systems and some multiple-occupancy buildings. See Section 230. Service conductors supplying a building or structure can't pass through any other building or structure. Installation of the service is explained in Chapter 7.

In urban areas, you usually make a written request to the utility for electrical service. An electrical load estimating sheet and a sample request form are shown in Figures 6-25 and 6-26. Utility rules will differ from area to area. Some require only the estimating sheet. Others ask for a site plan or complete plan set. Your local utility will be happy to explain their procedures.

Motors
Requirements
If you handle much commercial or farm work, you'll often be working with motors.

As a general rule, every motor must have four devices:

1. Branch Circuit Overcurrent Protection. This device—either a fuse or a circuit breaker—protects the motor circuit conductors against short-circuits or ground faults.

2. Motor-running Overcurrent Protection. This protects the motor in case of overloads while running.

3. Controller. This starts and stops the motor. It's not required to open all the conductors.

4. Disconnecting Means. This actually disconnects the motor by opening all conductors. You may hear this called an isolating switch.

Not all motors in all circuits will need all 4 devices. For example, a stationary motor rated at 1/8 hp or less can use the same branch circuit protective device as the controller. A portable motor 1/3 hp or less can use its plug-receptacle connection. Motor requirements depend on:
• The horsepower and ampacity of the motor
• The number of motors in the circuit
• The type of motor

• The way the motor is connected in the circuit
• The means of starting the motor

Generally, the larger the motor, the more likely all four devices will be needed.

Sometimes you can combine one or more of these protective devices. For example, all four might be found in one piece of equipment: In this case, the motor would be on a separate circuit. The circuit has a manually operated switch with time-delay fuses. The switch is located at the service panel at the start of the motor circuit, in sight of the motor. The switch is both the disconnect and the controller. The time-delay fuses furnish both branch circuit and motor-running overcurrent protection.

Other combinations are possible. Another type of motor circuit may have the branch circuit overcurrent protection combined with the disconnect. The motor-running overcurrent protection is combined with the controller.

Equipment manufacturers supply a large number of motor combination devices. Your job is to recommend the combination that's most economical, easiest to install and meets code requirements.

The following sections outline motor installations. For most applications you'll have to check details in NEC Section 430. Manufacturers' catalogs and their representatives are a good source of information on installation practice.

When planning motor installations, use a work sheet like Figures 6-27 and 6-28. Table 6-29 will also be helpful.

Line of Sight Rule
The NEC requires that a motor and its driven machinery be within sight of its Controller. This means that the motor must be visible from the Controller *and* within 50 feet of it. If the distance is greater than 50 feet, it's considered out of sight even if it really isn't. The Disconnecting Means must be in sight of the Controller (OSHA limits distances to 15 feet).

Exceptions to the rule are allowed if the Controller or Disconnect can be locked in the open position. Or, if a manually operated switch to prevent starting the motor is in sight of the motor (See NEC Sections 430 and 422).

Motor Circuit Conductors
Branch circuit conductors to a single motor should be rated at 125% of the full-load current of the motor. Take the full-current load values from NEC tables at the end of Section 430. Nameplate values can't be used to determine conductor ampacity,

Electrical Load Estimate

Project _____

Location _____

Owner _____

Architect _____

Electrical Loads	KW Connect	% Demand
Lighting	_____	_____
Receptacles	_____	_____
Appliances	_____	_____
Small Power	_____	_____
Other	_____	_____
Air Conditioning	_____	_____
Electrical Space Heating	_____	_____
Water Heaters	_____	_____
Elevators	_____	_____
Other	_____	_____
KW Demand	_____	_____
Largest Motor HP	_____	_____

Building: New_____ Remodel_____ Addition_____ Other_____

Type of Service: Primary_____ Secondary_____ Overhead_____ Underground_____

Service Requirements
Entrance Size Amps   _____

Entrance Wire Size   _____

Voltage   _____

Phase   _____

Wires   _____

Load Estimate Sheet
Figure 6-25

_____, 19____

Madison Gas and Electric Company
P. O. Box 1231
Madison, Wisconsin 53701

Gentlemen: Attention: _____

   In accordance with your electric service rules on file with the Public Service Commission of Wisconsin, we hereby request 120/208 volts, three phase, four wire electric service for the following new building:

Name of owner: _____

Name of architect or engineer: _____

Address or location of building: _____

_____

   The connected load and estimated total 15-minute kw. demands for this building are as follows:

Lighting Equipment	**Connected Load kw.**	**Estimated demand**
Lighting: _____	_____	_____
Receptacles: _____	_____	_____
Small power: _____	_____	_____
Power equipment		
Air conditioning: _____	_____	_____
_____	_____	_____
Other: _____	_____	_____
_____	_____	_____

Total connected load and estimated demand—
lighting and power   _____   _____

   We further understand that we are to provide Madison Gas and Electric Company with an acceptable space for the transformer installation. We are enclosing a site plan and electrical plans to facilitate a prompt reply to our request.

   We also request that any correspondence or negotiations regarding acceptable transformer installation location can be directed to _____

_____

Yours very truly,

_____

Utility Service Request
Figure 6-26

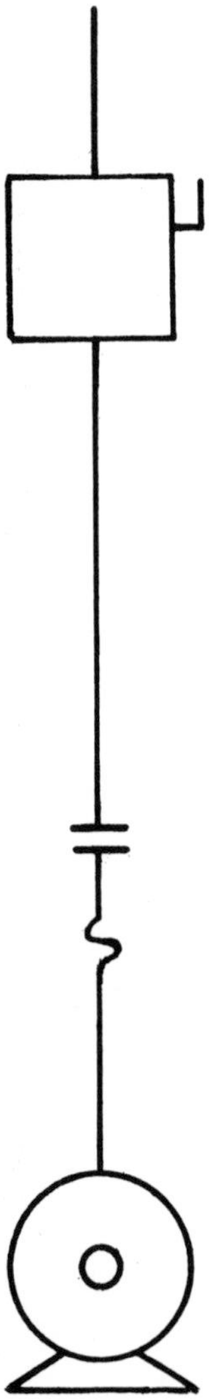

Switch size with one time fuses   _____ampere

Fuse size (one-time fuses)   _____amperes

Switch size with **Fusetron**
Dual-element fuses or **low-peak**
Dual-element fuses   _____ampere

Fusetron dual-element fuses or
Low-Peak dual-element fuses   _____amperes

Wire size _____ AWG

Conduit size _____ inch

Starter size _____

Running overcurrent rating   _____amperes

Motor horsepower _____ H.P.

Phase_____

Volts _____

Code letter on motor _____

Motor Branch Circuit Worksheet
Figure 6-27

switch amp ratings or short circuit/ground fault protection. But nameplate current ratings can be used to find motor overload protection. They are also used for full-current loads of specialty motors with low speeds and high torque.

For conductors supplying several motors in one circuit, find the wire size by taking 125% of the full-load current of the largest motor and 100% of all others. This is true for motors of different or equal size.

Some exceptions to these ratings are allowed for the motors not operating full-time. Check the NEC Table in 430-B.

Sometimes the branch or feeder circuit supplies one or more motors plus a heating and lighting load. In this case, wire size is found by adding the heating and lighting loads to the motor load.

Notice that long conductors to a motor may have to be increased in size to prevent voltage drop larger than 2%.

Motor Branch Circuit Protection

Branch circuit overcurrent devices for motors have to be rated to carry the full-load current shown in the tables at the end of NEC Section 430. The device must provide enough time delay so that the motor can start and accelerate to normal operating speed without tripping the breaker.

These devices must also be able to carry the starting current which can run as high as 250 to 300% greater than the full-load current. Maximum ratings for branch circuit protective devices for

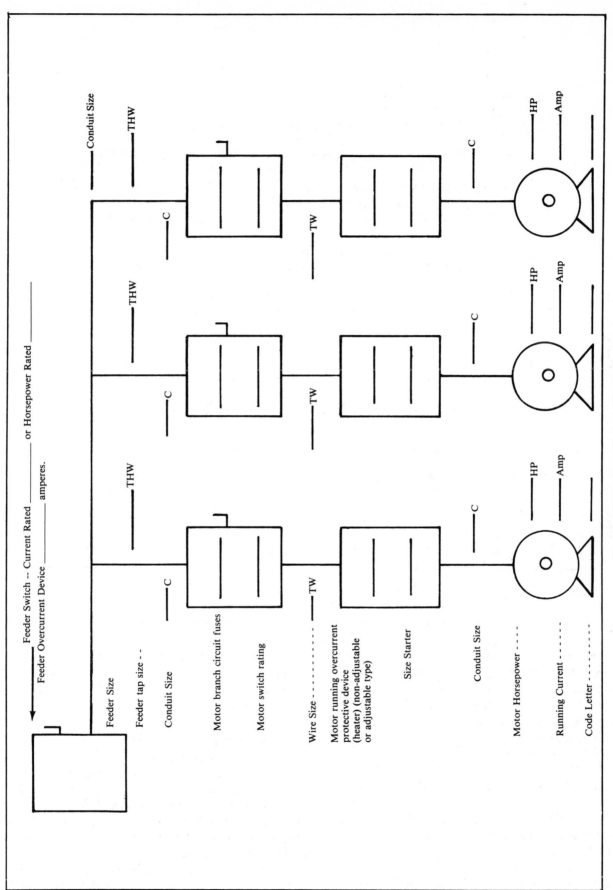

Group Motor Installation Worksheet
Figure 6-28

Motor Branch Circuit Sizing Chart
Induction Type — Single Phase

HP	Full Load VA.	120V Amp	208V Amp	240V Amp	Brk. Size Amp	Switch & Fuse Amps Fusetrons	Switch & Fuse Amps Regular	Wire Size	Cond	Max. Ft. 2% Volt. Drop
1/6	506	4.2	---	---	15	30-7	30-15	2-No. 12	1/2"	375
1/4	677	5.5	---	---	15	30-10	30-15	2-No. 12	1/2"	290
1/3	828	6.9	---	---	15	30-12	30-20	2-No. 12	1/2"	230
1/2	1125	8.4	---	---	20	30-15	30-20	2-No. 12	1/2"	220
3/4	1588	---	7.3	6.6	20	30-20	30-30	2-No. 12	1/2"	200
1	1840	---	8.9	7.7	20	30-20	30-30	2-No. 12	1/2"	175
1 1/2	2300	---	11.1	9.6	20	30-20	30-30	2-No. 12	1/2"	150
2	2760	---	13.3	11.5	30	30-20	60-40	2-No. 12	1/2"	125
3	3910	---	18.8	16.3	40	30-30	60-60	2-No. 10	3/4"	130
5	6440	---	31.0	26.8	70	60-45	100-90	2-No. 8	3/4"	130
7 1/2	9120	---	44.2	38.9	90	60-60	200-125	2-No. 6	1"	125
10	11500	---	55.1	48.1	100	100-80	100-150	2-No. 4	1 1/4"	160

Motor Branch Circuit Sizing
Table 6-29

motors are given in NEC Table 430-152. The NEC allows smaller fuses and breakers than these if the devices have time-delay features suitable for the particular motor. See NEC Section 430-D, *Size of Fuseholder Exception.*

If you have two or more motors rated at 1 horsepower or less and if neither draws more than 6 amps, NEC Section 430-D lets you put them both on a single circuit. But this circuit must be protected at not over 20 amps at 120 volts, or 15 amps at 600 volts or less. Each motor must have its own running overcurrent protection.

Two or more motors of any rating may be connected to one branch circuit. See NEC Section 430-D. Each must have motor overcurrent protection. The motor controllers and overcurrent devices must be a type approved for group installation. The branch circuit must be protected as described above for the largest motor, plus 100% of the full-load currents of the other motors. However, the protection can't be greater than 400% of the smallest motor rating.

Motor-running Overcurrent Protection

Select motor-running overcurrent protection devices by the motor nameplate amp rating.

For motors over 1 horsepower or automatically starting motors that are 1 horsepower or less, automatic start protection must be provided by *one* of the following:

1. A separate overcurrent device that responds to motor current. It must be rated or set at not over 125% of the full-load current for motors marked for 40 degrees C rise or having a service factor not less than 1.15. For all other motors, it must be set at not more than 115%.

2. A protective device built on the motor that responds to current or current and temperature change.

Some specialty motors are considered properly protected if they are part of an approved assembly. An example of this would be a motor on an oil-burning furnace which is controlled through the safety combustions controls of the unit. Protection of this kind must be shown on the nameplate of the assembly.

Other very small motors and motors with low torque have windings with enough impedance to prevent the motor from overheating if it should stall. An electric clock motor is this type. Motors such as these are also considered properly protected.

Motors of 1 horsepower or less that are not permanently installed, without automatic start, are considered protected by the branch circuit overcur-

rent protection. A separate motor overcurrent device isn't needed. Any motor not in sight of its controller must be protected in the same way as a motor with automatic start.

Suppose you're selecting a motor overcurrent protector. You calculate the value needed but can't find a standard size that meets your need. What then? You have to use the next higher size. But note that it can't be larger than:

1. 140% of full-load current for motors either with a temperature rise of 40 degrees C or sealed compressor motors; or

2. 130% for all other types.

Some motors use thermal devices to protect against overloads. A thermal detector alone won't necessarily protect the motor against shorts or overloads. The NEC requires protection from a fuse or circuit breaker. See Section 430-C on thermal cutouts and overload relays.

Motor Controller

Controllers are usually called starters. In general a starter must have a horsepower rating not less than that of the motor it controls. Some exceptions are allowed:

1. For stationary motors of 2 horsepower or less, a general-use switch with an amp rating of twice the full-load current of the motor may be used. On AC circuits, general use AC-only snap switches are acceptable. The full load current rating of the motor can't be more than 80% of the amp rating of the switch. Use "T" rated switches designed for high in-rush current.

2. A circuit breaker rated in amps may serve as a starter. If it's also used as the motor overcurrent protection, it must have a rating appropriate for the motor. If the motor has other overcurrent protection, the breaker is used simply as the starter. In this case, refer to the section in this chapter on motor branch circuit overcurrent protection for the breaker rating.

3. Some motors don't need a separate controller. Stationary motors of 1/8 horsepower or less that are normally left running can use the fuse or breaker as the controller. Some small motors are designed so that they can't be damaged by overload or starting failure. These can use the fuse or breaker as the controller. Portable motors 1/3 horsepower or less can use their plug and receptacle as the controller.

A controller doesn't have to open all the ungrounded conductors to a motor unless it's also the disconnecting means. This permits the use of single-pole pressure switches and similar devices in 230 volt circuits.

The controller must be in sight of the motor and within 50 feet. If this isn't possible, you have two choices: You can use a controller that can be locked in the open position, or you can install within sight of the motor a manually operated switch that can prevent starting.

It should be noted here that while the NEC requires a controller or switch within 50 feet of the motor, OSHA requires that it be within 15 feet.

Magnetic Starters

One type of starter uses a magnet to open and close the circuit. When this type is used, it must have a manually operated disconnect switch between it and the power source. This is needed in case the magnetic circuit fails to open. The manual switch can shut off the power even if the starter fails.

The manual switch could be the motor circuit switch at the point where the circuit begins. If practical, it could also be the disconnecting means. Or, a separate manual controller may be needed. Whatever switch you use, keep it within sight of the motor.

Disconnecting Means

The disconnecting means in the motor circuit must disconnect both the motor and controller from all ungrounded conductors. No pole can operate independently. The disconnecting means must be in sight of the controller unless it can be locked open. It can be in the same enclosure as the controller.

The disconnect must be a switch rated in horsepower or a circuit breaker. It must be readily accessible. If it's a switch, the horsepower rating must not be less than the motor. If a breaker is used, the amp rating must be at least 115% of the full-load current rating of the motor. Ratings for combination load disconnects—two or more motors or motors and resistance loads—are in NEC Section 430-H under *Ampere Rating and Interrupting Capacity*.

For single motors, the following exceptions apply:

• For stationary motors 1/8 horsepower or less, the branch circuit overprotection device can be used as the disconnect. In other words, a fuse can be removed or a circuit breaker can be opened.

• For motors of 2 horsepower or less and 300 volts or less, a general-use switch is acceptable. It must have an amp rating at least twice the full-load current of the motor.

• For portable motors, the plug and receptacle can be used as the disconnect.

• A general use switch is OK for torque motors. The switch must be rated at 115% of the motor nameplate current.

• A general use switch or an isolating switch can be used for stationary motors rated at more than 40 horsepower DC or 100 horsepower AC. The switch must be plainly marked "Do not operate under load."

A switch or circuit breaker can serve as both the disconnecting means and the controller. But it must meet the rating requirements of the controller and must open all the ungrounded conductors, as required for the disconnect. It must also be protected by a circuit overcurrent device which will open all the ungrounded conductors to the switch or circuit breaker itself. Usually this would be the branch circuit fuse or breaker. Finally, the switch or breaker must be the type that's thrown by applying a hand to a lever or handle.

Disconnecting means are required for each motor in a circuit. The following exceptions apply:

• Where a number of motors drive several parts of a single machine and one disconnect would serve all the motors.

• Where a group of motors is connected to a single branch circuit as described in NEC Section 430-D.

• Where a group of motors is in a single room within sight of the disconnecting means.

Types of Motor-protective Equipment

Motor-protective equipment falls into two general types—*fuses* and *thermal-overload devices*. These devices are usually included with a controller.

Fuses are usually found with a manually operated controller. A common example is the typical safety switch with lever-operated blade contacts and cartridge fuses. However, the fuse-type also includes a small fuse receptacle and toggle switch which fits into an ordinary household-size switch box.

Where a switch and fuses are used for motor control and protection, any type of switch with twice the motor amp rating can be used *for motors 2 horsepower and less*. An ordinary safety switch with the proper size time-delay fuses will serve the purpose. For a small motor such as a ventilating fan, a simple toggle switch with the proper current rating would work. Time-delay fusing would have to be somewhere in the circuit.

For motors over 2 horsepower, the safety switch must have a rating in horsepower equal to the nameplate rating of the motor. Single-phase motors of this type require cartridge-type time-delay fuses. Single phase motors of 2 horsepower or less can use plug-type fuses unless prohibited by local regulations.

Thermal-overload devices operate on heat and current. They are used with two types of controller or switch—the *manual* and the *magnetic*.

The manual type is made up of a toggle or push button operated switch. The toggle or push button is mechanically connected to the switching mechanism. A heater selected for the specific motor current is arranged to trip a latch in overload conditions. A spring then opens the switch. The speed of the tripping is proportional to the size of the overload. In other words, the heavier the motor overload, the quicker the heater will trip the switch.

The mechanism can't be reset until the heating device has cooled. Resetting is done manually, unless an automatic device is built in.

This type of controller has a positive action. It's mechanically linked between the toggle or push button and the switch. Because of this, it's sometimes used as the disconnecting means as well as the motor control. However, since the heater can't react fast enough when short circuits take place, fuses or breakers must be located in the circuit ahead of the control.

The magnetic type of switch or control is made up of a switch operated by a magnetic coil. This coil is also called a holding coil. Current to operate the coil usually comes from the supply circuit to the motor. In some cases it can come from a separate supply.

The current to the holding coil may be controlled by either a push-button, a single-pole switch or an automatic switch such as a float or pressure switch. Where automatic control is used but manual switching is also desired, a switch called a selector switch or a hand-off-automatic switch is used in the control circuit.

In some systems, hand control is all that's needed. But it may be necessary to start and stop the motor from several locations. Push button control can be used, with as many stop-start buttons as are needed installed at control points. These push buttons use only the small current needed to operate the holding coil. They can be located a considerable distance from the controller holding the coil.

Because there's no mechanical control between the "pilot" device used for control and the switch mechanism itself, a separate disconnect must be placed in the circuit ahead of the magnetic starter.

With a selector-switch or other automatic control, any interruption in the current supply will cause the motor to stop. When service is restored, the motor will start automatically if the holding coil circuit is still in the "On" position.

Where push-button control is used, service interruption will also stop the motor. But it won't restart when service is restored. You have to push the start button. This type of control is useful on machinery which might cause damage if it were started while unattended.

Like the manually operated thermal-overload control, the magnetic switch also uses a heater which opens in case of motor overload. In this case the heater operates a contact which interrupts the flow of current to the holding coil. The holding coil then opens the switch, as it would in any situation where its current is stopped. As with the manually operated thermal-overload, the heater must cool before it can be reset.

Motor controllers with both the manual and magnetic thermal-overloads are made in a number of sizes:

Size 00 are for fractional horsepower motors used on 240-volt and smaller circuits. These small thermal switches will fit into an ordinary switch box. They use manual control.

Size 0 uses both magnetic and manual control. It can be used for any motor up to and including 1½ horsepower in single phase or 3 horsepower in three phase on 240-volt circuits.

Size 1 is available in both manual and magnetic. This size is used for motors up to and including 3 horsepower in single-phase or 7½ horsepower in three phase on 240-volt circuits.

Size 1½ has been developed for use on 5 horsepower, single phase motors. It's not available for three phase motors but comes in the magnetic type for all kinds of use. Manually operated controllers are available for some applications.

Size 2 is available only with the magnetic type of controller. It's used for motors up to and including 7½ horsepower in single phase, or 15 horsepower in three phase on 240-volt circuits.

Larger sizes of magnetic controllers are available, but the sizes just described are most common.

Magnetic controllers can also be used to start and stop motors in certain sequences. This can be useful for conveyor and farm feeder systems. Time-delay starting of motors is another useful refinement done through a motor control. More information on controllers and their application is available from manufacturers of motor control equipment and their service representatives.

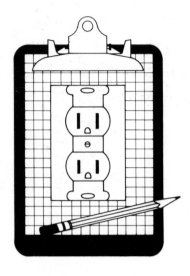

Chapter 7:

Installation for New Work

This chapter explains electrical work in new construction. The next chapter looks at the ways remodeling jobs are different.

Neither this chapter nor the next is a substitute for an apprentice program or several years of on-the-job training. But we'll hit most of the high points. Even experienced electricians will find information here that they can use.

Before starting to explain installation, we should cover one important point: Any part of the job, no matter how well done, will be done for nothing if it doesn't meet owner and inspector approval. Moreover, it will cost you money if it can be shown it was your fault and you have to re-do it.

If the plans are complete, if the specifications are clear, if the code is precise, and if you follow the plans, specs, and code exactly, you may never have a problem. But few jobs fit into that category. Too many electrical contractors have lost money on too many jobs because they began work without knowing exactly what was required. It isn't your fault that the plans, codes or specifications weren't clear. But it's always your fault if you go ahead with the job without getting clear instructions.

Don't add yourself to the list of electrical contractors who have lost money doing what they thought was right rather than finding out exactly what was required.

Here's how to avoid tear-outs at your expense: *Resolve every question before you begin.* Identify trouble spots and get specific guidance, in writing, from the owner, architect, general contractor and building inspector. See Chapter 11 for more information on revisions and change orders.

The owner, architect, and general contractor are obligated to provide specific instructions you can follow exactly. On many jobs they will meet periodically with all the subcontractors. Don't improvise. Set up more meetings with them if necessary to get all the details. That's not being over-cautious. It's just good professional practice.

Once it's clear what the owner, architect and general contractor want, resolve every code problem you can identify. Experienced electrical contractors know that there are three parts of every code question: *where, what and how. Where* can *what* material be used and *how* does it have to be installed? Forgetting any one of these questions may be a very expensive omission.

The first step is the *where*. Decide if the installation is approved for the proposed location. Suppose a plan shows an outlet on a garage wall. Can wiring be run under the concrete slab, saving on material costs?

Where often depends on the *what* question. Wire can go under the slab, but maybe a certain type of cable must be used. What kind of cable is needed? Answers about *what* are usually easy to find in the code. NEC tables explain what kinds of wire and equipment can be buried, embedded and exposed to moisture.

The last step is to determine *how* the installation must be done. Must the wire be in conduit if it's placed in the slab? How high must an outlet be from the slab? "How" questions are answered either in the NEC code or by your local inspector.

You also have to consider labor and material cost and job scheduling, of course. But these aren't the main questions at this point. Some types of electrical material, properly installed, can be placed almost anywhere.

If the question is service installation, ask the power company. Their recommendation will comply both with the NEC and local rules.

When questions come up, try to work with the inspector *who will be giving final approval.* A quick phone call may answer where, what and how in a few seconds. But remember that all installation approval, whether simple or complicated, involves all three questions. Make sure you have all three answers before doing the work.

One final note. Most installations need a visual inspection before approval. Don't make the mistake of burying or covering work before approval is granted.

Roughing-in and Finishing

Installation of interior wiring is generally divided into two major divisions: *roughing-in* and *finish work.* Roughing-in is the installation of outlet boxes, cable, wire and conduit. Finishing is the installation of the switches, receptacles, covers, fixtures and completion of the service. Between these two work periods other trades install wallboard, flooring and trim.

The first step in roughing-in is mounting outlet boxes on joists and studs directly or with special brackets. For concealed work, all boxes must be installed with the forward edge or plaster ring flush with the finished wall surfaces. If the boxes are used in a conduit system, they're attached loosely until the conduit is installed and tightened.

Have your foreman or crew leader mark the location of all boxes on studs and joists. See Figure 7-1.

Stud Markings Show Box Mounting Height
Figure 7-1

Roughing-In Cable Wiring
Figure 7-2

Next, drill or cut passage holes through studs and joists for cable or conduit. See Figure 7-2. Work goes faster if one person does all the drilling and cutting. A follow-up person then installs the cable or conduit. When this work is complete, have your foreman or crew leader make a visual inspection to be sure work done complies with the plans.

The final roughing-in step in a conduit system is pulling wires through the conduit. This can be delayed to become the first step in the finishing. But doing it during rough-in reduces the chance of damage to finished walls and ceilings.

The first step in finishing is splicing conductors in junction boxes. Follow the color coding. Then make connections between switches, receptacles, fixtures and conductors.

Next, attach devices and cover plates to the boxes.

Then connect the circuits and service entrance cable and install overcurrent protection devices (breakers or fuses) and metering equipment.

The final step is to "go hot" and check all receptacles with a test lamp or probe. Operate all switches. Connect a load to all circuits to be sure they are wired and protected correctly.

On large jobs you may be able to increase labor efficiency by re-arranging certain roughing-in and finishing steps. But stay in control of the job. Too much material installed too early can be a mistake.

Wire and Cable

Exposed wiring is a term used to describe an old type of installation where single insulated conductors were spaced apart and mounted on insulators. Except for a few industrial applications, exposed wiring is no longer used. Almost all new work falls into a classification called *concealed wiring*. Concealed wiring is run inside of walls, ceilings and below floors. Conduit, armored cable and types NM and NMC cable may be used, depending on the application.

Wire Installation
Most wire is run in conduit. For short runs with few wires, the conductors can be paired and pushed through the conduit from box to box. Longer runs or runs with several bends must be pulled with a fish wire.

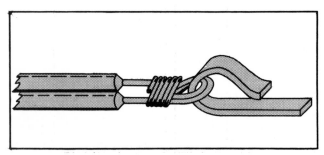

Fish Tape With Wires Attached
Figure 7-3

The fish wire is pushed through the conduit from one end to the other. Strip about three inches of insulation from the end of the conductors and wrap the bare wire around the hook on the end of the fish wire. See Figure 7-3. Then tape the splice. Taping prevents damage to the interior of the conduit and strengthens the splice.

For safety and efficiency, the wire-pulling team is usually two tradesmen. One person pulls the wires through while the other feeds them into the conduit. This prevents snagging and insulation damage. Special wire lubricants are available that ease the wire-pulling. Talcum powder also works well. Rub the lubricant or powder on the wires as they pass into the conduit.

On long or complicated runs, wire pulling may have to be done in sections between junction boxes. The same procedures apply, but it's going to take longer to pull each circuit.

Splice wire at the end of runs with a pigtail splice and wire nut or crimp connector. Wire splices are not allowed in the conduit.

Be sure wires are correctly color-coded before pulling. A two-wire circuit will use black and white; a three-wire circuit uses black, white and red.

Cable Installation
Nonmetallic sheathed cable (NM and NMC) can be used in concealed wiring where permitted by code. Usually it's installed in holes drilled through joists and studs or on the sides of joist and studs. See Figure 7-4. Cable must be supported by straps every 4½ feet and within 12 inches of each box unless it is run through holes.

In unfinished basements, it's sometimes convenient to run the cable at an angle to the joists. It can be supported directly on the edge of the joists if it is at least 2 No. 6 wires or 3 No. 8's. If smaller size cable is used, it must be installed through holes in the joists or along boards nailed to the joists. Some local codes will not allow any exposed cable in overhead locations. Check with the inspector for details.

If cable is installed across the top of a floor joist, it must be protected by guard strips at least as high as the cable. These are wooden boards nailed to the joist on either side of the cable. In a location not normally used, such as an attic or crawl space, guard strips are needed only within six feet of the entrance. Concealed cable installations should not be made near baseboards, door and window casing or other locations where they may be damaged by building nails. If thermal insulation is installed around the cable, it should be noncombustible and nonconductive.

To prevent damage to the sheathing on nonmetallic cable, the minimum allowable bend radius is five times the cable diameter.

Cable can be run in approved exposed locations both inside and outside the building if it follows the surface of the building finish or is on running boards. Where the cable is likely to receive physical contact, protect it with conduit, pipes, guard strips or by some other means.

Exposed cable passing through a floor must be enclosed in metal pipe or rigid conduit to at least six inches above the floor. Cable exposed to weather or wetness must have an approved sheathing.

You can't splice cable except in a junction or outlet box. Where a splice or connection must be made, strip at least 8 inches of cable covering from the insulated wires. See Figure 7-5. A cable stripper is best for this work, but a knife will do if you avoid cutting the conductors. After stripping and removing the insulated paper, fasten a connector to the cable and slip it through the knockout hole of the box. The locknut secures the cable from the inside. See Figure 7-6.

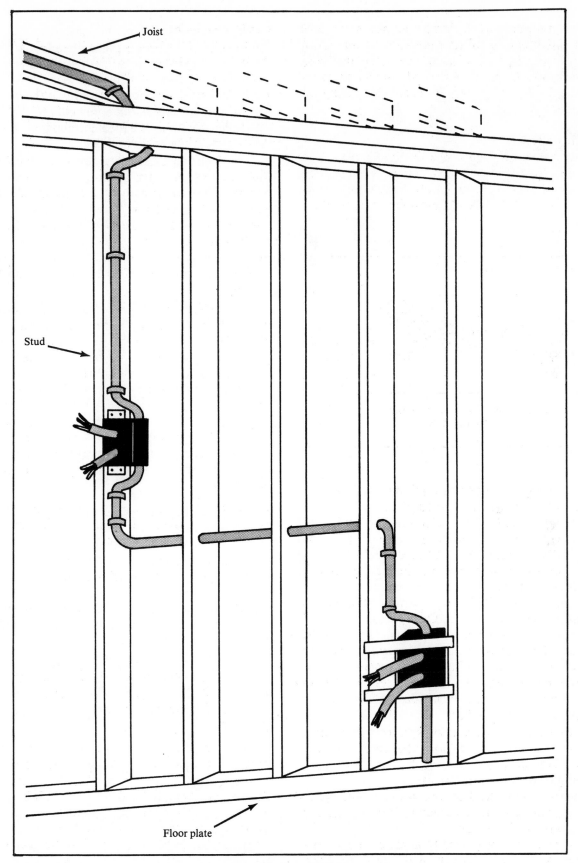

Typical Cable Installation
Figure 7-4

Number 14-3 Wire W/Gr Cable With Wires Stripped
Figure 7-5

Conduit

Your local code or the NEC tells you where conduit is required. Chapter 4 of this book will also help answer conduit questions.

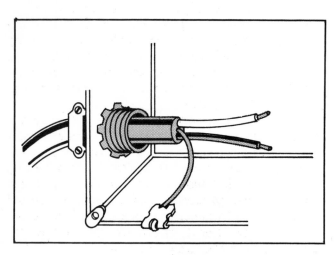

Cable-Outlet Box Installation
Figure 7-6

Rigid Conduit

When bending rigid conduit, be careful not to collapse the tube or reduce its internal diameter. The curve radius at the inner edge of the bend must not be less than about eight times its internal diameter. Chapter 4 shows the exact dimensions. No more than four 90 degree bends are permitted in a run between two boxes. No more than three 90 degree bends should be made in any ten foot length.

There are several acceptable ways to make bends in rigid conduit. Factory-made bends save time but may cost more than making the bend on site. They come in standard dimensions such as 45 degrees and 90 degrees. Straight conduit is then simply cut to the correct length and connected to the factory-made bend with a threaded coupling. Entrance ells are a type of factory-made bend.

You'll probably want to use a mechanical conduit bender on a larger job for fields bends. For conduit over 3/4'' diameter a mechanical bender is essential. Conduit 3/4'' and less can be bent by hand with a hickey. Figure 7-7 shows how to make a hand bend:

1. Suppose the bend needed is 90 degrees at 20 inches from the conduit end. Put a mark on the conduit at this point.

2. Slip the hickey on the conduit so its back edge is placed two inches ahead of the mark, and bend the conduit about 25 degrees.

3. Move the hickey back two more inches and bend the conduit about 45 degrees.

4. Move the hickey back another inch and bend the conduit to about 70 degrees.

5. Move the hickey back one inch again and make the final bend to 90 degrees.

Rigid conduit is always bent with a series of small bends as the hickey is moved along the pipe. It's better to use several small bends rather than one or two larger bends. Bending the rigid conduit too much in any one place will kink it.

For odd-shaped bends, draw the shape you need on the floor in chalk. Match the conduit to your drawing as you bend.

Here's a tip that can save many mistakes: Leave the ends on the bend longer than needed. Cutting to length later is easy. Sure, you'll have some waste, but it will save the loss of the entire length many times.

Rigid conduit has the same diameter as standard water pipe. It can be cut with a hacksaw or pipe cutter and is threaded with standard threading tools and dies. Use cutting oil while threading and cutting. Remember to ream each end after it's cut and before it's threaded. This removes the sharp inside edge.

On large jobs you'll use power tools for the cutting and threading.

Before installing any conduit, inspect it for foreign matter inside and thread condition on the outside.

Make every conduit run as straight and direct as possible. When several lines of conduit are run side by side, install adjacent lengths at the same time if possible. This usually saves time.

Conduit is supported by straps or hangers. See Chapter 4 for maximum distances. On wood surfaces, nails or wood screws can be used to support the straps. Masonry and brick surfaces need anchors to hold screws which support the straps. On a metal surface, drill holes and tap them to accept machine screws.

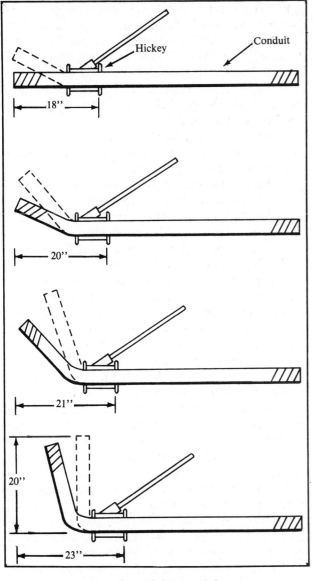

Bending Rigid Conduit
Figure 7-7

For certain applications, devices such as nail-ups and clamp-backs speed up the supporting work. Notches cut in studs or joist are enough to support horizontal runs.

Either a threaded end or a factory connector can be used to connect the conduit at junction and outlet boxes. A factory connector is needed if the code requires a rain-tight or concrete-tight joint.

In either case, run a bowed locknut over the thread with teeth facing the conduit end. Some locknuts have special features such as a sealing surface or grounding screw. Run the conduit or connector through the box knockout. This is easier if the box is mounted loosely until the conduit is installed. Screw a bushing tightly on the conduit end. The bushing has a smooth inner surface to protect the wires when they are pulled through. Finally, tighten the locknut outside until its teeth dig into the metal of the box.

Some factory connectors have a built-in bushing. In this case, use the collar or locknut of the fitting on the outside of the box. Another locknut secures the fitting inside. Tightening the locknuts is important. The conduit can't act as a ground if doesn't make good contact with the box.

Installations will be different, depending on what type of connectors are on hand. In some cases, the connector will butt to the box and only a single inside locknut is needed.

Rigid Nonmetallic Conduit
PVC conduit (see Figure 7-8) is quick and easy to install because it's lighter and can be cut with a hacksaw. Bends are made either with factory elbows or by field bending, where a small, portable oven is used to heat the conduit to a temperature that permits bending. Factory elbows are connected with couplings.

Make coupling connections with PVC cement. The cement allows the pieces to be joined and moved for a few seconds. Then it sets very tight. Connections to boxes are made either with a factory threaded conduit end or with factory connections. When gluing, be sure the conduit is fully inserted all the way into the connector or coupling. Avoid breathing the cement vapors. Don't have the container near an open flame or spark.

Make box connections to PVC the same way as rigid metallic conduit. But it isn't necessary to tighten locknuts as much because the conduit isn't a ground. In fact, overtightening locknuts on PVC conduit will strip the threads.

PVC connectors don't need bushings to protect wires where they enter the boxes.

Thin-wall Conduit
Thin-wall conduit is lighter and easier to work with than rigid metallic conduit. It has a thinner wall but the same interior diameter and cross-sectional area as rigid conduit.

PVC Conduit Before Concrete Slab is Poured
Figure 7-8

As with rigid conduit, be careful to avoid kinking or reducing its inside diameter when you make a bend. The radius of any field bend must not be less than about eight times the diameter of the conduit. Not more than four 90 degree bends are permitted in one run. See Chapter 4 for precise requirements for each size.

A thin-wall bender is used to make field bends. It has a cast-steel head with a pipe handle about four feet long. There's a bender for each size of conduit. Use only the bender made for the size you're bending.

There's a right way to use a thin-wall bender: Lay the conduit on a firm, level surface. Hook the bender head under the conduit. Apply steady pressure by hand and by stepping down on the step of the bender until you've formed the right angle.

When bending any length, maintain the *take-up distance*. This is the bend radius at the inside of the curve. For 1/2-inch conduit, the take-up distance is five inches, 3/4-inch conduit uses 6 inches, and 1-inch conduit needs 8 inches. Subtract the take-up distance from the stub length and mark that point. Put the index mark on the bender at this point and

make the bend. Figure 7-9 shows how to bend an 11'' stub end.

Thin-wall conduit may be cut with a hacksaw or special thin-wall cutter. Ream the sharp interior edge to prevent wire damage. Thin-wall conduit is never threaded. Connections are made with compression or screw-type couplings and connectors. Fittings which are secured by indentations in the tube are also available. These require a special indenting tool.

Thin-wall is placed and supported just like rigid conduit. Box connections are also similar. As with rigid, be sure all connections are tight so there's a grounding path for the system.

Flexible Conduit and Armored Cable
Cut flexible conduit (flex) and armored cable (BX) with a hacksaw run at an angle to the conduit. See Figure 7-10. Cut through one section of the flex and then twist until it breaks. If wires are inside, be careful not to cut them. Put a bushing inside the end of the conduit after the cut is made. This provides a smooth surface at the end of the run and also strengthens the conduit for the connector.

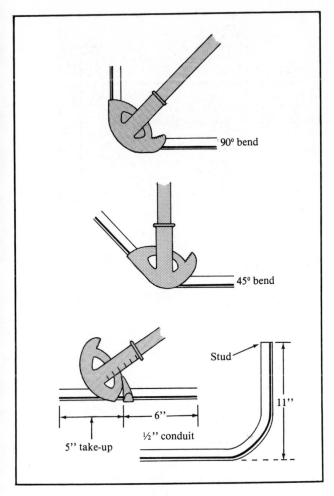

Bending Thin-Wall Conduit
Figure 7-9

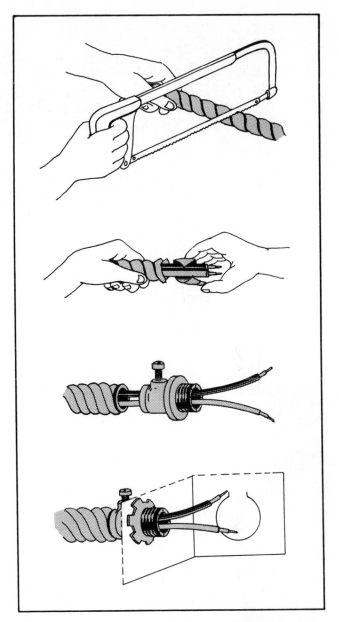

Installing Flexible Conduit
Figure 7-10

Make box connections the same way as with rigid conduit.

Some flexible cable and conduit has a bond wire bent back against the flex. At each junction, fasten the wire to the screw of the connector. This guarantees a good ground path.

Restrictions on the length of flexible conduit runs and support needed are explained in Chapter 4.

Switching

Figure 7-11 shows a basic switching circuit. Figures 7-12A through 7-12E show other possibilities. Fixture switching is simpler if you understand the following principles:

• Neutral and ground wires are never connected to a switch.

• Each fixture must have a switch loop.

The loop is made up of two parts: a hot wire running from the source to the switch, called the *switch leg*, and the wire running from the switch to the fixture. This is called the *return leg* and is hot or cold, depending on the position of the switch.

For ganged switches, one source wire may be the switch leg for two or more switch loops. For three- and four-way switches, the return leg may actually be made up of two or more wires. But no matter how complicated the wiring becomes, it still relies on a basic switch loop to control the fixture.

In any two-wire circuit, it's common practice to use a black and a white wire. The black wire is the hot wire. The white wire is the neutral. However, in

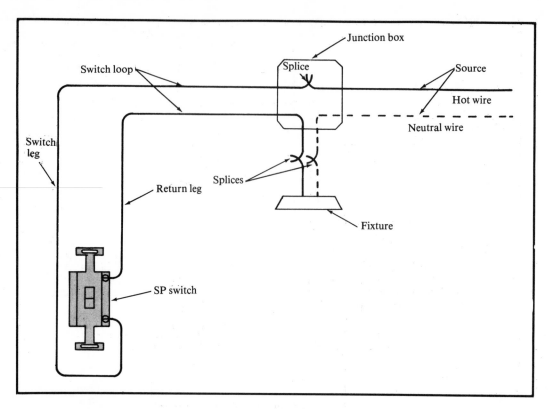

Basic Switching Circuit
Figure 7-11

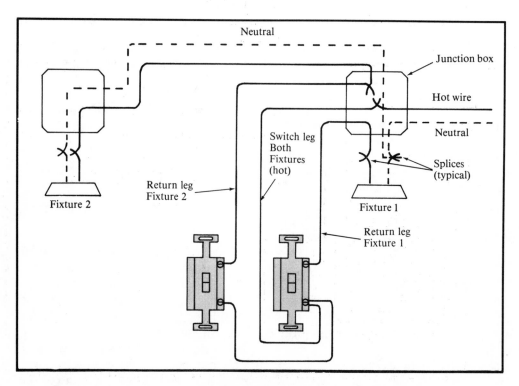

Two-Gang Switch for Separate Control of Two Fixtures
Figure 7-12A

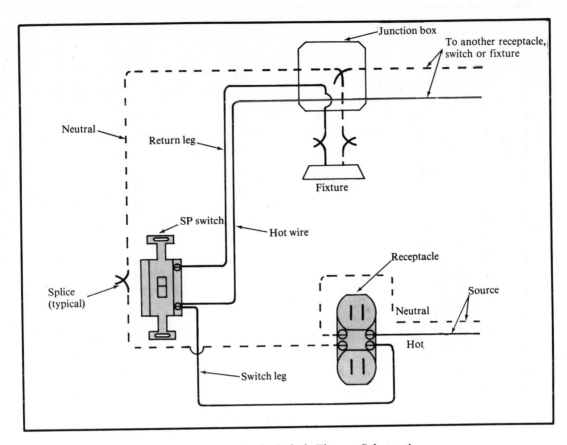

Typical Receptacle-Switch-Fixture Schematic
Figure 7-12B

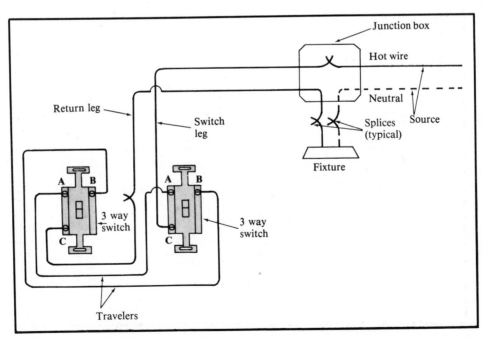

Three-Way Switching with Fixture Beyond Switches
Figure 7-12C

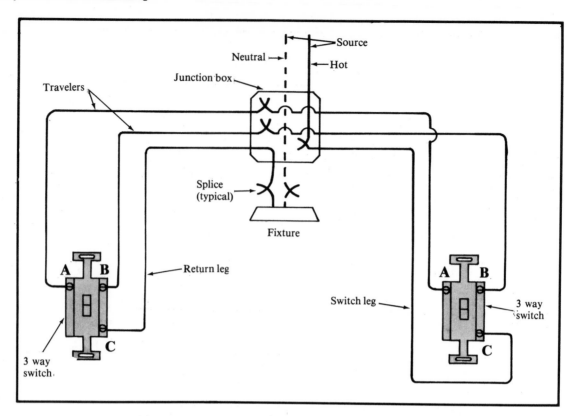

Three-Way Switching with Fixture Between Switches
Figure 7-12D

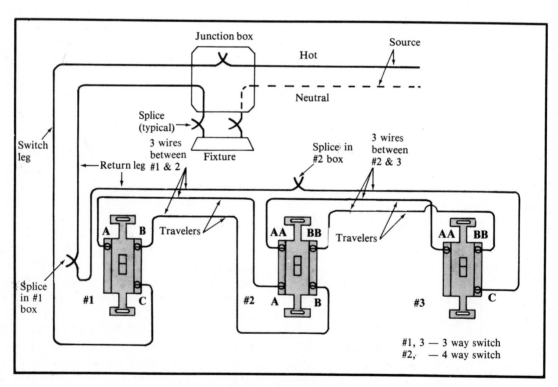

Four-Way Switching
Figure 7-12E

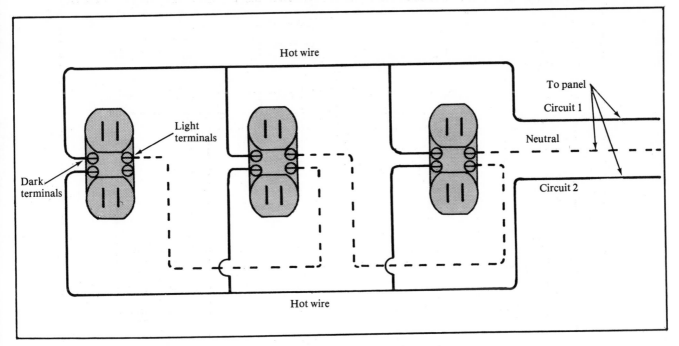

Split-Circuit Receptacles Schematic
Figure 7-13

a typical switching circuit, the neutral wire is needed only up to the fixture. Because of this, it is acceptable for the white wire between the switch and fixture to become a hot wire. It becomes the return leg of the switch loop. To show this, you must paint the white wire black at all switch and fixture connections.

Don't forget that the white wire is now a hot wire. It is no longer the neutral. It cannot be wired to the neutral wire in the circuit. Also, don't make the mistake of painting the white neutral wire attached to the fixture.

Three- and four-way switches need more than two current-carrying wires in their circuits. Besides the black and white wires, a third wire is used. This is usually red. A red wire is also needed when the white conductor serves its usual purpose as the neutral in a multi-way switch circuit. This happens when a receptacle outlet is installed in the circuit along with the switch and fixture. The white neutral wire is pulled through the switch junction and on to the receptacle.

To avoid confusion in the drawings, the neutral is shown as a dotted line. Current-carrying wires are drawn solid. They may be black, white (with painted ends), or red, depending on the type of installation.

Whatever the wire color, the hot wire coming from the source should always be wired to the dark

colored terminal on the switch. This is generally the lower terminal when the switch is installed correctly—with the "off" position pointing down. The light colored terminals are used for the control legs of the circuit. Drawings usually show switch connections on the front of the switch for clarity. In practice, most switch connections are made on the side.

Study the electrical plans carefully so you know how many wires are needed to switch the circuit. In conduit, the right number and color wire can be pulled together. For cable installations, you may need three- or four-wire cable in certain parts of the circuit. Make a quick field sketch of more complex circuits to avoid mistakes.

Split-circuits and Switched Receptacles
In kitchens, workshops and similar areas, it's good practice to use split circuits for the convenience receptacles. This makes overloading the circuit less likely. *But check your local code* before making this type of installation.

Here's how to install a split circuit: Run two hot wires of the amperage needed from two different breakers in the panel. Run a neutral wire from the neutral bar. See Figure 7-13. The neutral is connected to the light-colored terminals on both the upper and lower receptacles and serves all receptacles in the circuit. One hot wire is connected to

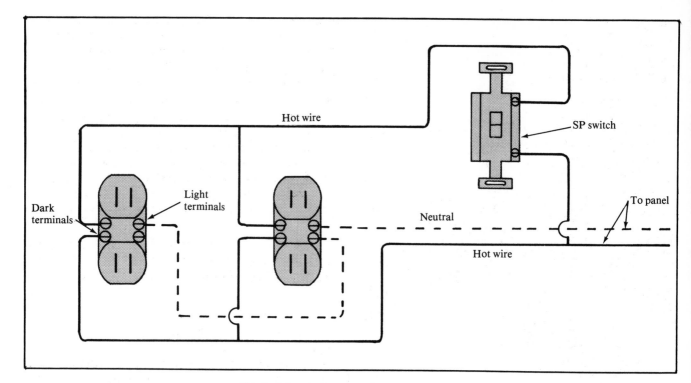

Single Circuit Receptacles - Upper Half
Switch Controlled
Figure 7-14

the dark terminals of all the upper receptacles. The second hot wire connects to all the lower dark terminals. Normally both the upper and lower receptacles need only one hot and neutral connection to make them both hot because of a lug between them. Be sure to *break the lug* before installing the receptacles in a split circuit.

On some jobs the customer may want to control the receptacles on the circuit with a switch. To switch-control the receptacles in a split circuit, install a single-pole switch in either the upper or lower hot wire, depending on which half of the receptacle wants control. Be sure to check with your local inspector first, since switch-controlling split circuits is sometimes not permitted.

It's more likely that you'll run into a job where the plan or customer calls for switch-control of single-circuit receptacles. This is common in living rooms.

To switch the upper receptacles in a single circuit, the receptacles are wired in the same way as if for a split circuit. However, the upper hot wire will not run back to the circuit breaker. Instead, connect it to the hot wire running between the lower receptacles and the breaker. The switch is installed in the upper wire as usual. See Figure 7-14. Again, break the lugs.

To control both upper and lower halves of single-circuit receptacles, simply install a switch in the hot wire. No other wires are needed. However, check local codes, since often only one half of the receptacle can be switched.

Equipment Circuits
Equipment circuits usually have to be custom designed for the type of equipment being installed. Anything that draws high wattage, such as water heaters and motors, will use a separately fused and sometimes separately switched circuit from the service panel. See Figure 7-15.

Equipment circuit installations are described in Chapters 5 and 6. Your local code or your electrical utility may impose special requirements.

Plug-connected equipment such as ovens and larger air conditioners use one-position plugs. The plug can be inserted in the receptacle only one way. The plug slot configuration allows you to insert it in only the right size receptacle. But you have to select and install the correct receptacle configuration.

Any equipment that has exposed parts which could become hot when a current defect happens must be grounded. This applies for all circuits over 150 volts to ground and any equipment in a hazar-

A Fused Safety Switch for Air Conditioner
Figure 7-15

dous location. It also covers a wide variety of equipment in residences and other occupancies. For many jobs only a grounding conductor is needed in the equipment cord. This conductor connects with the conduit or circuit-grounding conductor. But be careful when installing the plug and receptacle that you don't confuse the grounding connections with the neutral connections.

Figure 7-16 shows common receptacle circuiting for equipment connections. (Figure 4-30 in Chapter 4 shows other configurations.) The letter G identifies the ground. Letter W is the neutral connection. Terminals X and Y are connected to hot wires.

Figure 7-17 shows wiring for double-pole switches in a 240-volt circuit. Be sure the wire and receptacle are adequate for the amperage.

Grounding
Types of Grounding

1. A *system ground* is the ground applied to a neutral wire. It limits the possibility of fire and shock by reducing one of the system wires to zero volts potential above ground. It protects the system against a high voltage surge such as a lightning strike. The system ground also is a path for the return of excess circuit voltage that is common in normal operation.

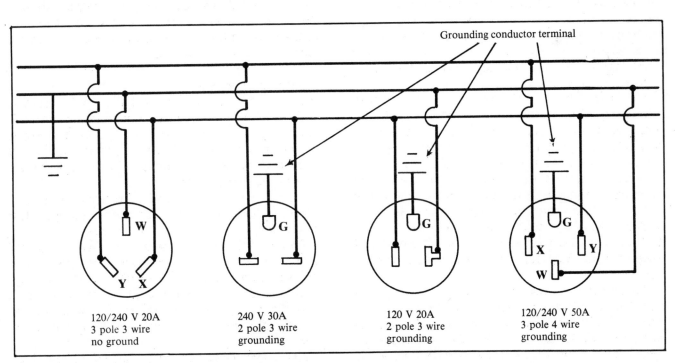

Equipment Receptacles Schematic
Figure 7-16

The neutral wire is color coded white. You may hear it called the *grounded* conductor.

2. An *equipment ground* is an extra ground in the system that's attached to equipment and receptacles. Its purpose is to prevent shock if the frames or bodies of appliances become charged due to a fault. The equipment ground wire may be either bare or have a green covering. It's called the *grounding* conductor.

Ground Installation

A *system ground* is made by connecting a No. 4 or larger braided copper wire to a cold water pipe or a

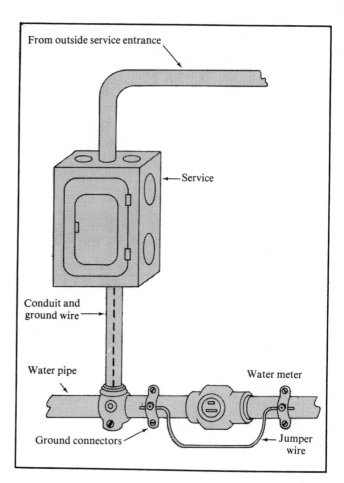

An Urban Residential System Ground
Figure 7-18

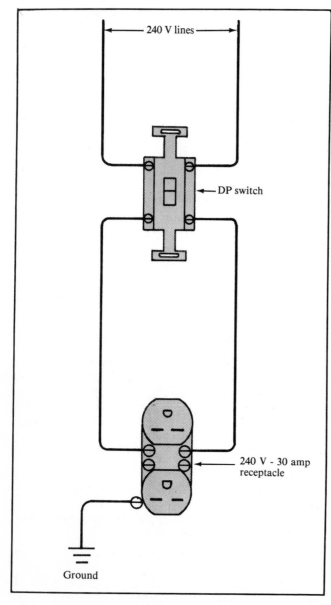

Double-Pole Switch in a 240 Volt Circuit
Figure 7-17

copper ground rod driven in the earth. The other end of the wire is connected to the neutral conductor in the service. Where the connection is made depends on power company rules for the installation.

The water pipe method was common for older urban residences. The ground wire usually was connected to the neutral bar in the service panel. This method is still used for some types of services. See Figure 7-18. It cannot be used with plastic water pipe.

You'll more likely find that power company and local codes will require a ground rod. The rod must be copper or copper-coated, a minimum of 5/8 inch in diameter and 8 feet long. Locate it at least 2 feet from any building and drive it at least 12 inches below the surface. Some power companies will ask for two rods. If you install the rod in an excavation, ask *before burying it*, if any visual inspection is needed.

Both underground and overhead services will have the ground wire connected at the service side

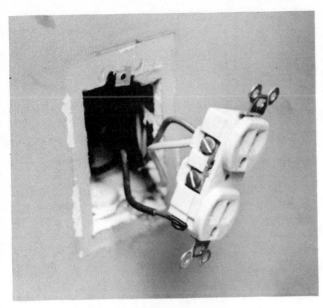

Receptacle Grounded with Pigtail
Figure 7-19

of the meter socket or the panel neutral bar. The other end is clamped to the rod. Be sure the rod is sanded clean where the wire is clamped.

Some codes require that the ground wire be run in conduit. It can't be in the same conduit with any service wires.

Most power companies have handbooks describing acceptable system ground installations. Also check with local inspectors.

Equipment grounding in metal conduit systems uses the conduit as the ground path. In cable and nonmetallic conduit systems, a grounding conductor is needed. With either method, plugs and receptacles use special slots, prongs and connections to tie the ground conductors together. The ground prong on the plug is connected to the metal frame of the equipment. The receptacle slot is connected to the conduit or grounding conductor which connects to the neutral strap inside the service panel.

The neutral strap is the common ground-connection point for grounding conductors or conduit, neutral wires, and the ground wire connected to earth. Conductor connections are made with screws. Conduit is grounded to the neutral strap with a ground bushing and a short bond wire.

In some systems the panel itself must be grounded. Most manufacturers supply a ground screw inside the panel for this purpose.

In a conduit system, circuit receptacles must be wired to their boxes to guarantee a good ground path. This is done with a ground screw and pigtail.

The screw is inserted into a threaded hole in the box. Its pigtail is connected to the green-colored hex screw on the receptacle. See Figure 7-19.

For systems using grounding conductors, the receptacles are grounded to the boxes in the same way. This prevents the box from becoming hot in case of circuit failure. You may prefer to run the grounding conductor itself from the receptacle ground screw to the box screw, eliminating the pigtail. Loop the wire and you don't have to do any cutting.

Figures 7-20A, 7-20B and 7-20C show three common methods of equipment grounding.

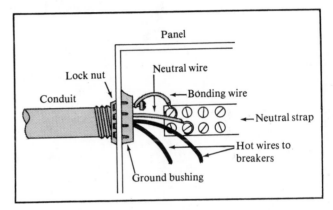

Grounding Conduit at the Service Panel
Figure 7-20A

Service Entrance
Your power company will have a booklet which describes their requirements for service entrance to their customers. The utility may also ask you to complete a current demand report and identify the date service is needed. Chapter 6 shows typical forms. If your local code requires permits and inspections, these must be completed before the utility will turn on the power.

Overhead Installations
In most locations, wiring up to the service entrance head above the watt-hour meter is done by the utility. You'll put in the rest of the service, except for the meter. You'll mount the meter socket, but the meter itself is supplied by the utility.

Service can be installed either with entrance cable or conduit. Most areas now require conduit. Your code will explain whether thin-wall or rigid must be used.

All service installations begin with an entrance cap. This is also called a rain cap or service cap.

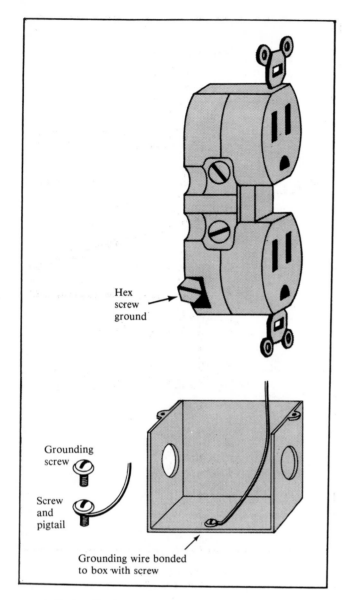

Grounding Receptacle and Box Equipment
Figure 7-20B

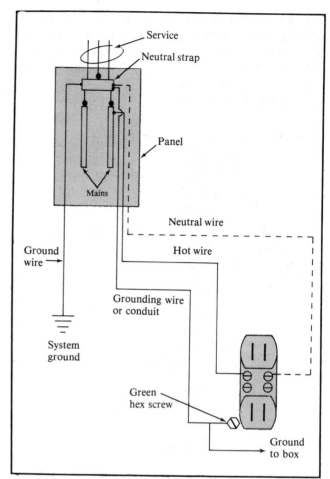

Receptacle Grounding Schematic
Figure 7-20C

The cap should be mounted so the service drop will be at least ten feet above finished grade where it reaches the building. It must not be closer than three feet to windows, porches or other wall openings.

The service drop is connected to the side of the building by a service rack or insulators. The NEC requires that the entrance head be installed slightly higher than the top insulator. This keeps water from running down the wire to the entrance head.

The service entrance wires must extend out of the entrance head for 36 inches. Strap the conduit or entrance cable every four feet down to the top meter socket. This socket should be weatherproof.

The cable or conduit then continues from the bottom or side of the meter socket to where it enters the building at an entrance ell or sill plate. The ell is used for conduit installations. It has a removable plate to help in threading the wires. A sill plate is used with entrance cable to make the opening in the building watertight.

Check with your utility for the correct meter socket mounting height. Generally it must be placed at eye level—between 54 inches and 72 inches. Use an approved socket. Its location—inside or outside the building—should also meet utility approval. Provide at least three feet of unobstructed space in front of the meter.

Service wires can usually be pushed through the conduit by hand.

In some areas, entrance wires must be color coded. Otherwise, where accessible only by authorized persons, they can be black. Sometimes you'll have to mark the neutral with white paint.

The neutral wire will connect to the middle connections of the meter socket and the neutral strap in the panel. Hot wires will connect to the two outer terminals of the meter socket and the main buses in the panel. Figure 7-21 shows a typical overhead service entrance.

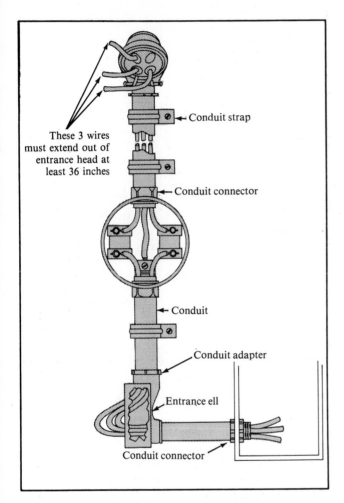

Typical Overhead Service Entrance
Figure 7-21

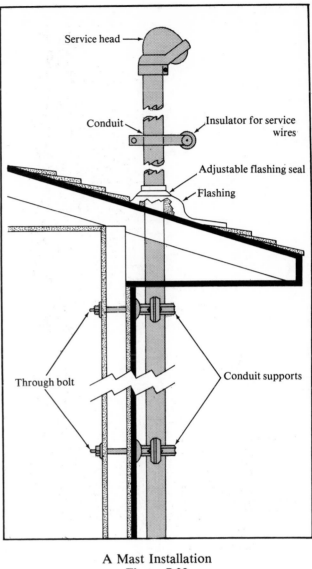

A Mast Installation
Figure 7-22

In some cases, the building may not be high enough to give the service drop the correct ground clearance. Then a service mast is needed. Figure 7-22 shows a mast installation. Anchor the mast securely. It has to carry the weight and weather loads placed on the service drop.

Underground Installations
Many utility companies provide underground service materials up to a specified distance at no charge to their customers. After that, the customer must pay at some specified rate per additional foot.

Usually the utility owns and maintains the underground service lateral and the meter socket. You have to install the meter socket and rigid or PVC conduit. Wiring and a suitable ground system is also your responsibility.

Start by installing the meter socket at the right height. A length of conduit—from 2 inches to 4 inches in diameter, depending on service amperage—is run from the meter socket to 18 to 24 inches below grade. Both ends of this conduit must be protected by a bushing to avoid damage to the wires. Strap the conduit every four feet. Generally, the system ground wire is not allowed in the conduit.

Check connections in the meter socket to be sure they are large enough for the underground cable.

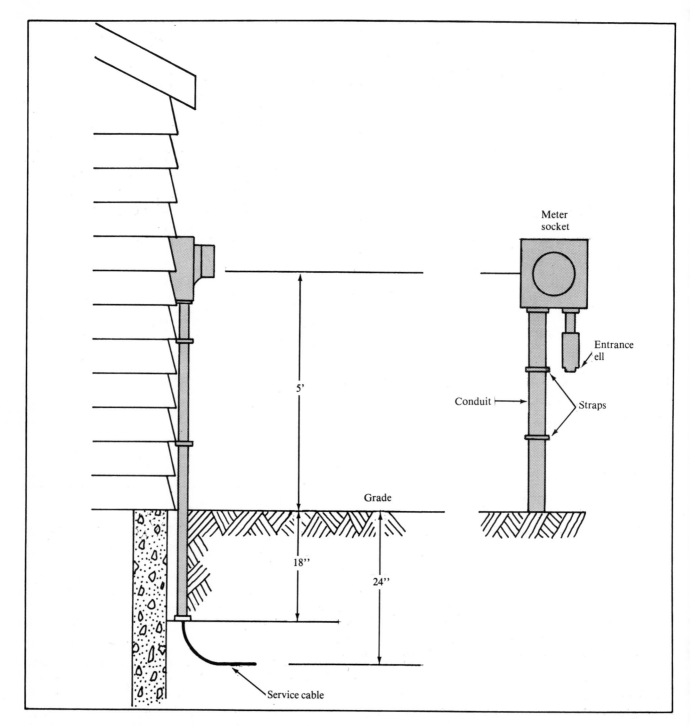

Typical Underground Service
Figure 7-23

Be especially alert to this potential problem when wiring the building service to the meter: The building service must connect to the *lower set of taps* on the meter socket. The underground cable enters the socket from the bottom. Some electricians are tempted to attach service entrance wire on the lower connections of the meter. Meters, however, will run in only one direction. Incoming current connections must be made at the top of the meter socket.

Figure 7-23 shows a residential underground installation. Figure 7-24 shows meter wiring for

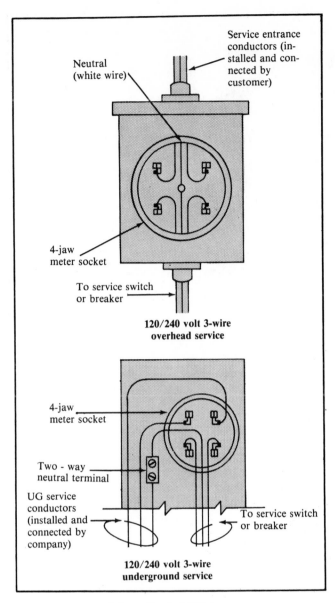

Meter Socket Wiring
Figure 7-24

Meter Installation for Underground Service
Figure 7-25

overhead and underground services to a four-jaw meter socket. A completed meter installation is shown in Figure 7-25.

Underground Wiring

The best service connection is underground. So too, there are plenty of good reasons to run feeder and branch circuits underground where the code allows:

1. It eliminates the chance of wires coming down in wind or ice storms.

2. It does away with restricted clearances where lines cross driveways or other obstacles.

3. It improves the appearances of the property.

4. It removes the danger of striking exposed wires with machinery.

5. It allows the utility to locate transformers and metering equipment closer to the load center without the hazards of high voltage lines.

6. Replacement of wires or cable run in duct or conduit is easier than for an overhead installation.

The one big disadvantage to underground wiring is the higher cost of installation. Weigh the extra cost against these advantages.

Two types of conductors are generally used for underground installation—USE Cable and UF Cable.

Type USE is available in single and three-wire types. It may be used without fuses or breakers for service entrance. When protected with fuses and breakers, it may be used for feeder and branch circuits. USE is acceptable for direct burial. It can also be installed in underground duct or conduit. Its outer sheath is moisture resistant but it must be protected by conduit when used above ground on a pole or building. Since the outer sheath is not flame

A Meter Bank/Main Switch for an Apartment. Wire
Trough and Service Cable at the Left
Figure 7-26

resistant, it must be enclosed in conduit whenever it is used inside a building.

Type UF Cable is available in single, two- or three-wire types either with or without ground. UF is *not* approved for service entrance work. With fuses or breakers, it can be used for feeder and branch circuits.

UF is approved for direct burial and can be installed in underground ducts and conduit. Insulation on individual conductors must be both moisture- and flame-resistant. UF multiple-conductor cables can also be used inside buildings. When it is run outside, on the building exterior or a pole, it must be protected by conduit. UF cannot be used exposed to weather unless it is specifically designed for this application. It's generally available in No. 14 through No. 6 AWG.

Individual wires with type TW, RW, RHW, RH-RW, THW, RUW or THWN insulation can be installed in underground *conduit* runs.

Be sure any conduit you install underground is approved for underground work. And be sure you lay it correctly. Check Sections 345 to 352 of the NEC. Also, your utility company may impose special requirements for services.

Where conduit is used to protect Types USE or UF cable on a pole or the side of a building, make sure the conduit runs to from 12" to 24" below ground level and at least 8' above the ground. Higher clearances may be required, depending on the type of installation.

Conduit entering a building below grade must be sealed to prevent water from entering the foundation. Avoid cable entry through foundation walls wherever possible.

Trenches for underground installation should be at least 24" deep. Some codes require a 42" depth. Go deeper if you think the soil will be disturbed by cultivation or ditching. Where a single-conductor cable system is used, all conductors, including the neutral, must be in the same trench, duct or conduit. For direct-burial installation, put a layer of sand under and over the cable to protect it from damage when the trench is backfilled.

Replacing direct-buried cable is expensive. Be sure the conductor size is adequate for both present and future loads. It's even more expensive to replace buried conduit or duct. Make sure it's large enough to allow for future replacement of larger conductors.

Commercial and Apartment Service

Many types of meter-bank and meter-switch combinations are available for commercial and multiple-unit residential installations (Figure 7-26). Your choice should be based on economy, ease of

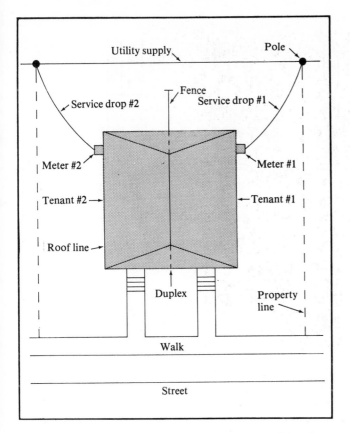

Typical Duplex Service - Separate Drops - Plan View
Figure 7-27

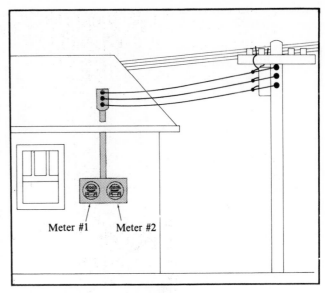

Typical 2-Unit Dwelling - Service Entrance
Separated at Meters
Figure 7-28

installation and the code. Of course, both equipment and installation methods must meet power company requirements.

Some installations use a pad-mounted transformer. Usually the utility supplies the transformer at their cost. The general contractor is responsible for the concrete base. Either you or the electrical utility will install wiring between the transformer and the building.

Check both your code and utility requirements before starting work. Be sure the materials you use are approved for the type of installation planned. And pay special attention to equipment and conductor clearance rules and burial depths.

Generally only one service drop and entrance is permitted to a building. Exceptions are allowed under NEC Section 230-A. For example, a building with a very large floor area may have two services. Another exception is when there is not enough space in a multiple-occupancy building for service equipment accessible to all the occupants. A separate service is also allowed if some emergency or stand-by system requires it. Multiple-use buildings may need separate services for different voltages, phases, controlled water heating or different rate charges.

Residential duplexes and townhouses will need separate drops and services when the tenants are completely separated. Sometimes a single drop is made and the service entrance wires for each unit are tapped from that drop. Where only a few units are involved, you may be able to provide the separate service entrances at the meter bank. See Figures 7-27 and 7-28. Your utility will have rules that cover this situation.

There must be some way to disconnect all conductors in the building from the service entrance conductors. Usually this is a switch. Be sure the switch is the right size, type and has the right markings. Codes usually require you to install the switch within 5' of where the service enters the building.

Where more than one building is on the same property and under the same management, each building must have a separate disconnect. Service wires can pass through one building to another under these conditions. Normally, service wires can't run through buildings.

For multiple-occupancy buildings, each occupant must have access to his disconnect switch. Mount all switches in a group and mark each to show the load served. The switches can be installed either inside or outside the building. But they must be located near the point of service entrance. Your power company will be able to recommend the best location.

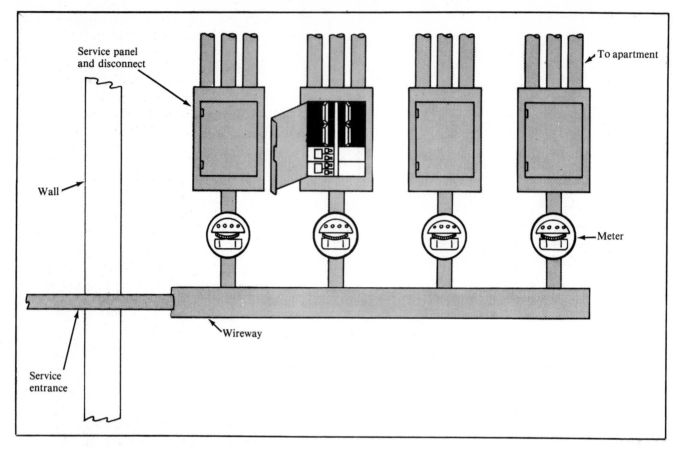

Typical Grouping - 4-Unit Building - Separate Service
Figure 7-29

Two or more separate sets of service entrance can be tapped from one service drop in multiple-occupancy buildings. Up to six disconnects, grouped and marked, can be tapped. When more than six disconnects are needed, a main disconnect switch must be installed between the service and the separate disconnect switches. See Figures 7-29 and 7-30.

Generally, service entrance conductors can't be spliced. But splicing is permitted at an approved location such as a wireway, when separate disconnects are needed. Clamped or bolted connections are also allowed in metering equipment enclosures. Other exceptions are listed in NEC Section 230-F.

You may want to draw up a detailed plan and materials list of the service to sort out any questions that come up. Once the plan is completed, get a preliminary approval from the inspector.

Here's a checklist you should review before submitting the service entrance plan for approval:

1. Type of Service
   a. Overhead
   b. Underground
   c. Utility requirements
   d. Code requirements

2. Conductors and Conduit
   a. Location—correct clearances, depths and paths
   b. Approved size, insulation or coating
   c. Where required in installation

3. Equipment
   a. Approved size, type, markings

4. Installation
   a. Correct equipment location, order and clearance
   b. Approved ground system
   c. Approved splice locations

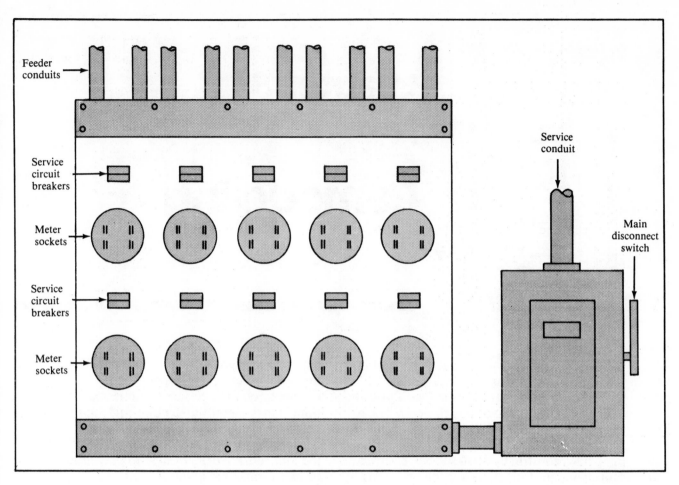

10-Unit Meter Bank with Main Disconnect Switch
Figure 7-30

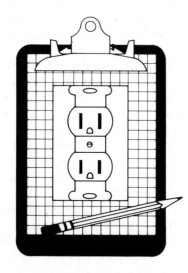

Chapter 8:

Remodeling

Some electrical contractors avoid remodeling, room additions and repair work. They feel it's more trouble than it's worth. This is fine if you're lucky enough to always have enough new work to keep you going. But in Europe and many parts of the U.S., there's been more repair and remodeling done than new construction lately. No electrical contractor can afford to ignore this big and growing market.

For many beginning electrical contractors, remodeling and repair is a good place to start. Especially if they're small, these jobs won't attract the big outfits that have the men and equipment on hand for a low bid.

And keep this in mind: although repair and remodeling work involves more problems and unknowns, it usually carries a higher markup for the alert, skilled electrical contracting firm that takes the time and trouble to build a reputation for doing it well.

If you handle primarily residential and small commercial jobs, include remodeling and repair of residential and commercial buildings in your scope of interest. If you prefer industrial and heavy commercial work, also bid remodeling and repair of industrial and commercial work. The only real mistake is limiting your scope to strictly new construction.

But remodeling, room additions and "rehab", as it's sometimes called, is different. You need specialized knowledge and skills. And that's what this chapter is about.

The most common rehab or remodeling job for electricians is up-grading the electrical system to handle bigger loads. If you know what you're doing, adding amperage capacity can be a quick and efficient job. Older installations may take more time, but generally ingenuity and knowledge can speed and simplify this type of work.

Planning is extremely important in all remodeling and rehab work. In new work you know what to expect and are familiar with work conditions. Every remodeling job includes unknowns and potential code problems that are hard to resolve until work is well under way. But careful investigation and a clear understanding with the owner of what your bid includes and excludes can avoid most unpleasant surprises.

Code Restrictions
The NEC requires that wiring done to an addition over 500 square feet meet all requirements for new work. That is, the types of circuits and their loads must be designed and calculated under the specifications in Section 220-A, *Lighting Loads For Listed Occupancies.* Loads for new circuits or extended circuits in previously wired dwellings are designed following the specifications in Section 220-A, *Other Loads—All Occupancies.*

Your local code may vary from this standard. Be sure to check with your inspector. Most counties and cities require permits and inspections for remodeling just as for new work.

When the scope of the work has been determined, follow the suggestions for design and planning that are outlined in Chapters 5 and 6 of this book.

Residences

Start by reviewing the existing wiring system. You need to know how much work is required. Note the type of building construction, service size and capacity, type of wiring used and existing circuits. Have the owner describe chronic circuit overloads and what new circuits are needed. Be alert for areas that may give you installation problems.

After recording these key facts, work up an estimate as explained in Chapter 11. If you have trouble estimating the time and materials needed because exact building construction can't be discovered, you'll be safer if you can get the job on a time-and-materials basis. With this method, the customer agrees to pay you the cost of materials plus your hourly rate for installing them.

Commercial Remodeling

In commercial remodeling, a job "walk-through" is essential. Most remodeling contracts or job specifications include a clause that makes all existing conditions your responsibility. This prevents you from disclaiming responsibility later if some condition on site makes the work more difficult.

Make your preliminary inspection with the remodeling plans in hand. These plans are your guide and should include a plan set for the existing building. If they don't, and if they're essential to your work, get them. Before making the inspection, study the plans to see what's expected and how existing conditions may affect your work.

In remodeling as in new construction, plans are essential. But in remodeling, the plans may not accurately depict the existing building. The designer or architect probably didn't take the trouble to show all existing conditions. He and his client are more concerned with what's to be built rather than what's there now. Besides, he may show existing conditions for general construction purposes which may not tell you what you need to know for electrical work.

Let's look at an example. The electrical designer indicates that conduit should pass through a wall at a certain point. From the plans you can't see any problem. But you discover during your "walk-thru" that the wall in question is 12 inches of poured concrete around a shaft that can't contain wiring. Noting these types of conditions is essential to turning a profit on the job.

Your job-site inspection crew should be two people, the foreman or crew leader who directs the work, and the estimator who will bid the job. Take notes and talk over special problems. Raise essential questions with the architect or general contractor.

Review the plans again after making the job-site inspection. Be alert to conditions assumed in the plans that don't really exist or that are different from what the designer shows. Here's the place where your good judgement and experience can really pay off.

You're under no obligation to inform anyone of what you learned from the job-site inspection. If you spotted conditions which can only be handled through a change order after the work begins, that's your competitive advantage. This extra work will almost certainly be done at your price on a non-competitive basis. You don't have to bid this extra work and may want to exclude it specifically to avoid any disagreement later.

Don't overlook the value of removed materials. The custom is that the general contractor has salvage rights unless the owner retains it. But on some jobs the old wire and fixtures may be of more value to you. The general contractor may give salvaged material to you just for the asking. Remember to ask!

After you're satisfied you've got all the essential information, make up your estimate and bid (Chapter 11). Be as accurate as possible, especially in the area of who's responsible for the different types of contracting work that'll be needed. Don't forget to study the specifications thoroughly. Approach the job as if it were new work but with the added "horse-sense" that you'll need for profitable remodeling.

Computing Loads for Existing Dwellings

Compute circuit loads for remodeling and rehab jobs the same way as in new work. But you have one additional chore. Besides the circuit loads, you have to calculate the additional load the existing service can carry.

Section 220-C of the NEC allows the use of actual kVA demand figures to determine added service loads if the following conditions are met:

1. The kVA data of the existing demand is available for at least one year.

2. The existing demand at 125% percent plus the new load does not exceed the service ampacity.

3. The service conductors and overcurrent protection are of sufficient size for the increase.

kVA figures will be available from the utility which has been supplying the service. If the planned remodeling will increase the power needed by more than about 50%, the utility should also be informed. They may have special rules which you should know about.

Once you know the existing power demand, calculate the new loads as was described in Chapter

6. Add the new load to the existing load to find if the service is large enough for proposed changes.

Section 220-C also allows an optional calculation for an existing dwelling being served by a 115/230 Volt, 3-wire, 60 amp service. This method is allowed for a 60 amp service *only*. Here's how to do the calculation:

Take the first 8 kW of load at 100% and the remaining load at 40%.

The load calculation must include lighting at 3 watts per square foot, 1500 watts for each 20 amp appliance circuit and the nameplate rating of ranges, cooking units and other appliances that are permanently connected or fixed in place.

If air-conditioning or electric space heating is to be installed, *one or the other* must be calculated at 100%.

Air-conditioning equipment at 100%

Central electric heating at 100%

Less than 4 separately controlled heating units at 100%

The first 8 kW of all other loads at 100% and the remainder of all other loads at 40%

In other words, new heating or air conditioning loads must be added at 100% demand factor over and above the original calculation.

Here's an example: You have an existing 1,000 square foot dwelling with a 12 kW electric range. The building also has two 20 amp appliance circuits and a 500 watt furnace circuit.

1,000 sq. ft. x 3 W/sq/ft	= 3,000 W
Two 20 amp appliance circuits 2 x 1,500 W	= 3,000 W
Electric Range	=12,000 W
Furnace circuit	= 500 W
Computed Load	=18,500 W

Demand Factor:
8,000 W at 100%	= 8,000 W
18,500 - 8,000 = 10,500 at 40%	= 4,200 W
Net Load	=12,200 W

By multiplying 60 amps times 230 volts, we find the service capacity at 13,800 watts. The net load for the existing service is 12,200 watts.

$$13,800 \text{ W}$$
$$-12,200 \text{ W}$$
$$\overline{1,600 \text{ W}} \text{ may be added.}$$

If an appliance such as air conditioning or electric heating is to be added, it must be taken at 100% demand. Therefore, it can't exceed a 1600 watt rating. If, however, the 1600 watts is to be used for equipment other than that listed at 100% demand, the 1600 watts is divided by 40%.

$$\frac{1600 \text{ W}}{.40} = 4,000 \text{ W}$$

Up to 4,000 watts of equipment can be added at 40% demand.

Adding On Circuits

Most panels are installed with enough capacity for additional circuits. It's standard practice to leave extra knock-outs in the panel for additional circuit breakers (Figure 8-1). Many times, circuit capacity can be increased just by pulling old wires from the conduit and installing new wiring. For cable systems, new cable can be added.

With older systems, or where a large number of new circuits are planned, it's easier to add another panel to the system. Feeder panels of this type are easy to install. They're tapped off the main service panel connections. Circuit wiring and breakers are installed as with any job. Remember to allow for additional circuits at a later date if the feeder has a large enough capacity. See Figure 8-2.

Check the service capacity before the feeder panel is installed. The feeder panel, conductors and its circuits must be the right size and ampacity. Make these calculations as explained for new work in Chapter 6. Also, refer to NEC Section 215 on feeders. Your local code may vary these requirements.

Surface Wiring

Where walls are not open during construction, surface metal raceway is a good choice. But surface raceway isn't allowed in all locations. Be sure to check with your inspector before doing the work.

Surface raceways are sometimes called wire moldings. They're similar to conduit. Wires are pulled through after the raceways are installed and connected to their boxes. Raceway is grounded like

Residential Panel with Extra Knockouts
Chapter 8-1

conduit. Matching receptacle and switch boxes are available for raceway so the finished installation looks complete and professional.

Raceway goes up fairly quickly. The channel is usually fairly flat and is available in neutral colors such as grey, brown, and ivory. But it's always going to be apparent on the wall and some property owners may not like this. Visibility is reduced if it's run close to baseboards and door trim. It can also be painted after installation to match the existing wall color. Figure 8-3 shows how surface raceway looks when run beside baseboard and trim.

Where approved by your local code, two- and three-wire exposed cable surface wiring systems can be used. This is common where appearance and wire protection are not important. Surface wiring devices such as switches and receptacles are available for surface cable installations. As with raceways, reduce visibility by running the cable next to baseboard and trim and by painting the finished installation.

Before installing an exposed cable system, check the NEC rules for that particular location. As a general rule, cable must be run flush against a surface. Where it runs in open air, such as across floor joists, it may need added protection.

Running Wires in Existing Walls and Ceilings

Usually it's too difficult to run rigid conduit or EMT in walls and ceilings when remodeling. Most codes permit the use of flexible conduit or plastic-sheathed cable in their place. Installation of cable or flex takes patience, ingenuity and care.

You'll need two or more fish wires or tapes for remodeling work. By working through existing outlets and drilled holes, the two fish wires can be used to pull wires where one fish wire alone might not work.

A small plumb bob will help you locate blocking in walls which would prevent wire passage. Magnetic stud finders are also useful.

A keyhole saw is essential in remodeling work. Also, long-shanked drill bits and rotary bits that cut large diameter holes are handy. Many types of boxes and receptacles are marketed especially for remodeling. Your supplier can tell you which ones will work best for a particular job.

Figures 8-4A through 8-4F show how to run wiring through walls, floors and ceilings. These details will be useful when doing any remodeling work. Keep them handy. They also illustrate how two fish tapes can be used to good advantage.

In many cases you'll have to remove baseboard or door trim. When finished, replace the trim to conceal the opening.

Before removing any trim, be sure a replacement is available. Many styles of wood trim, especially styles found in older homes, are no longer manufactured. If you damage a piece in remodeling, you may have to replace all trim in the room so that it matches.

Removing trim is quicker with the right tools. Use a good nail puller or small pry bar. Be careful when applying pressure that you don't slip and damage the wall surface. A rubber pad or folded cloth placed between the tool and the surface will save a lot of touch-up work.

When it is necessary to remove tongue-and-groove floor or wall boards, cut carefully with a circular saw set just to the thickness of the board. With careful handling, the saw can also be used to cut the tongue off the first board that's removed. This prevents damage to the board remaining in place. Use a wood chisel for notching or trimming the boards left in place. Before replacing floor boards and wall trim, be sure the cable is where it won't suffer nail damage.

Sometimes you'll have to cut into a wall to find existing wiring. Finding the exact position of what you're looking for may not be easy. Always start with a small pilot hole that can be patched easily if you're at the wrong spot. When your pilot hole is

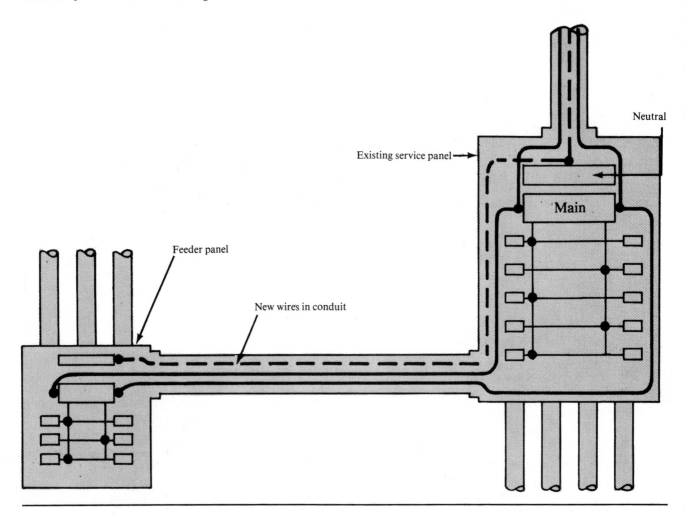

Typical Feeder Panel Installation
Figure 8-2

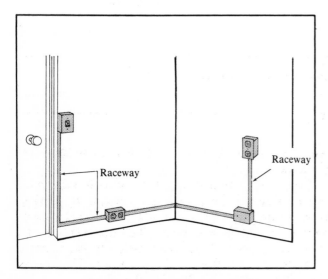

Raceways for Surface Wiring
Figure 8-3

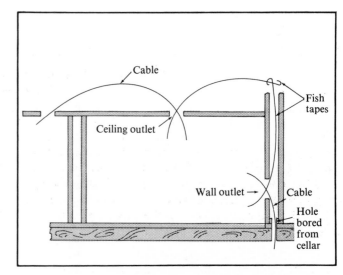

Typical Cable Installations - Using Fish Tapes
Figure 8-4A

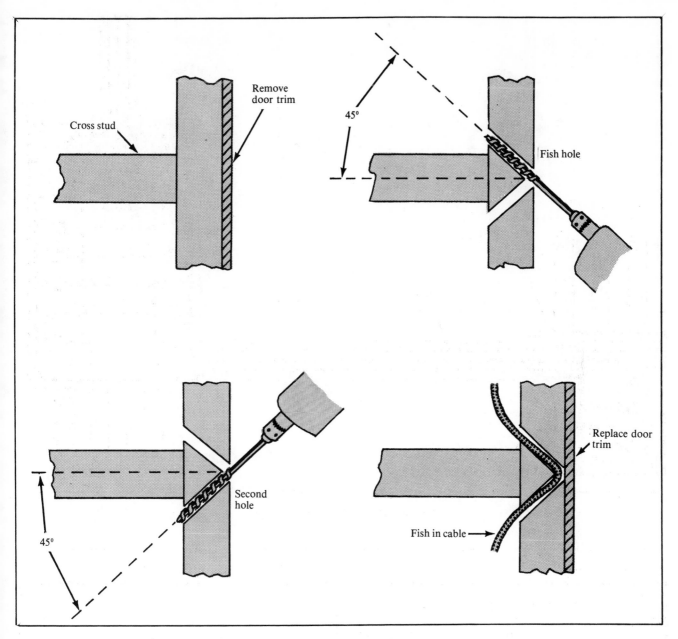

Typical Cable Installations - Bypassing a Cross Stud
Figure 8-4B

on target, then cut a larger hole. The same principle applies when you have to cut all the way through a wall. Start with a small drill that's long enough to penetrate the wall and cavity. See where the drill hole comes out on the other side. If the location is right, then cut the larger hole.

Troubleshooting

Sometimes you'll be called to locate a problem circuit. The circuit may be dead or continually short-out. The following basic steps will solve the problem most of the time:

The most common cause of circuit failure is overloading. If the circuit is protected by a plug fuse, a good sign of overloading will be the clear window in the blown fuse. The fuse window remains clear since an overload melts the metal strip at a slower rate than a short circuit. The lower temperature does not discolor the window.

For cartridge fuses and breakers, there are no visible clues. Check for overloading by counting

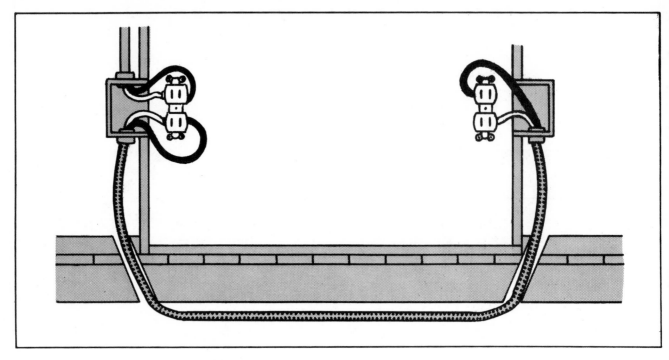

Typical Cable Installations - Outlet Installation
Using a Floor Joist
Figure 8-4C

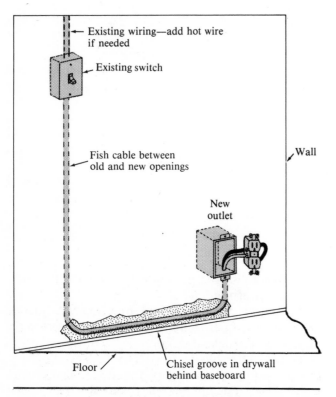

Typical Cable Installations - Installing a New
Receptacle From an Existing Switch
Figure 8-4D

the appliances and equipment of the circuit. The homeowner's explanation of the problem will offer good clues to the problem.

A temporary overload is common when a motor kicks on. Motor-starting current may be three times the operating current. When a motor causes the overload, check the fuses for proper size and type. Time-delay fuses should be in the circuit. If they aren't, check the motor rating and install the right size time-delay fuse.

Constant overloading will also cause circuit failure. A homeowner who has to shift appliances to prevent circuit failure needs additional circuits or larger capacity. There isn't much you can do about this type of overload except install new circuitry. Where one circuit is overloaded, others are probably in the same condition. An alert electrician can take this opportunity to explain the advantages of a remodeled, modern electrical system in that home.

A totally dead circuit or a circuit with intermittent losses of power is usually caused by mechanical failure. You won't face this problem very often. Since it's so rare, you'll suspect mechanical failure only after spending an hour or more looking for shorts and overloads.

If there has been a mechanical failure, check the circuits and their overprotection devices for loose

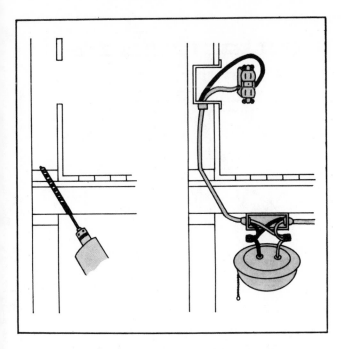

Typical Cable Installations - Running a Cable from an Existing Receptacle to a New Fixture in the Floor Below
Figure 8-4E

---

connections. A wire may have been broken. Loose wire screws and broken or bent prongs on breakers and fuse holders are the most common culprits. Don't forget the fuse itself. It may look good and still be defective. Check with an ohmmeter or replace with a fuse you know is good.

Sometimes the homeowner is at fault. He has replaced a plug fuse with a fuse that's too short. Or he may have inserted a cartridge fuse and not seated it properly. He may have mistakenly used an old fuse. When his problem still isn't solved, he calls the electrician, fearing a more serious problem. You may be mistakenly led into looking for a subtle problem without first checking the basics.

With any circuit problem, connections should be checked first.

Circuit failure is also caused by short-circuiting. With a plug fuse, the high temperatures and high current surge of a short will cause the window to discolor. In a breaker system, this high surge may cause a sparking around the breaker connections inside the panel.

A short circuit can be caused either in the circuit itself or in its equipment. To check the source of the problem, unplug or disconnect all equipment on the circuit. Then turn the breaker back on or replace the fuse. If the new fuse blows or the

breaker trips, the short is in the circuit wiring or fixtures.

Although a short in the wiring itself isn't impossible, usually the problem will be a defective switch or receptacle. A visual inspection of these devices may disclose the problem. If not, use an ohmmeter. Finally, check the wiring with an ohmmeter.

If the replacement fuse or circuit breaker doesn't blow or trip, it shows that an appliance is at fault. Turn each one back on, one at a time. The faulty one will cause the circuit to blow. Then replace or repair the appliance.

Occasionally, a fuse or breaker itself may be defective. Use a test lamp to check breakers and fuses in the panel. With the current on, hold one end of the tester against the hot connection. Hold the other end against the neutral wire in the neutral strap. The light should glow. Be sure to hold the tester only by its insulated probes. And be careful that the area where you are standing is dry.

A breaker with a loose or wobbly switch is probably bad. Replace it.

Making a Circuit Survey

A circuit survey may save time and mistakes when remodeling or handling an overload problem. Your survey will show where circuitry must be added to accommodate existing and new loads.

A circuit survey can be done on any blank sheet of paper. You'll need columns for circuit numbers, outlet locations and wattages. See Figure 8-5. If the service panel circuits are numbered correctly, the survey numbers will correspond.

Begin the survey by turning on all lights in the building. Be sure all the lamps are good. Turn the breaker off or remove the fuse for circuit No. 1. Go through the building and note the location of unlit lamps. Identify them opposite circuit number one on your survey form. Do the same for all receptacles. Use a night light or test lamp to determine if the receptacle is hot. Be sure the receptacles are not wall-switched. If they are, the switch should be on.

Continue with this procedure until all circuits have been switched off. Every receptacle and light should be listed on your form. Don't forget exterior fixtures and receptacles and mechanical equipment such as furnaces and pumps.

When all circuits have been identified, the wattage for each can be figured. For appliances, use the nameplate rating. For receptacles with a specific use, the wattage of the item plugged in can be used. Light fixture wattage is taken from the

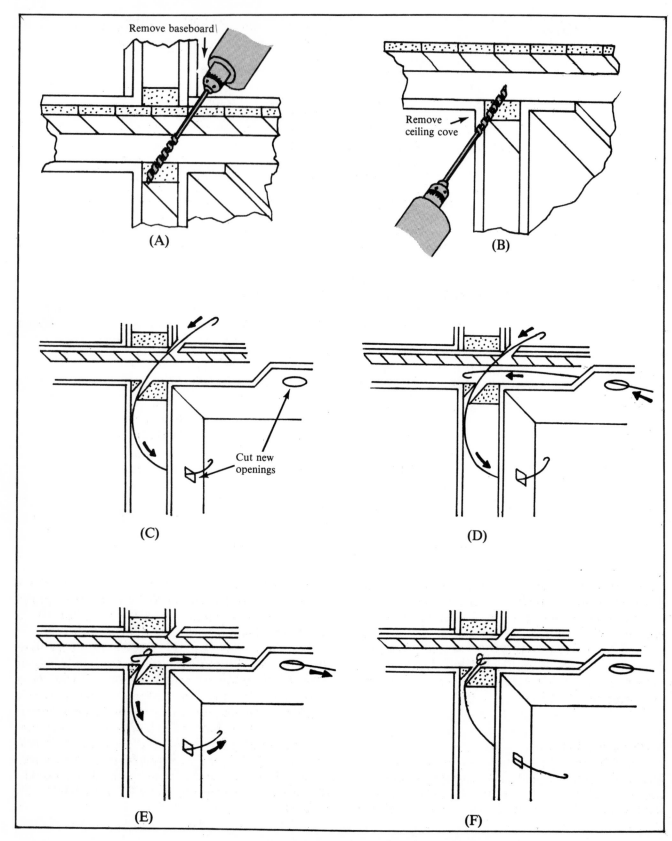

Typical Cable Installations - Using Two Fish Tapes for
New Work in an Existing Wall
Figure 8-4F

Circuit Number	Location	Outlet	Load #1	Load #2	Load #3	Load #4	Load #5
1	Living room	Ceiling light	150W	--	--	--	--
		Walls -					
		Lights	275W	--	--	--	--
		TV	300W	--	--	--	--
		Stereo	150W	--	--	--	
	Dining rm.	Ceiling light	100W	--	--	--	--
		Conv. rec.					
		3@ 180W	540W	--	--	--	--
2	Bathroom	Ceiling light	--	100W	--	--	--
		Mirror light	--	60W	--	--	--
		Conv. rec.	--	180W	--	--	--
	Bedroom #1	Ceiling light	--	75W	--	--	--
		Conv. rec.					
		3 @ 180W	--	540W	--	--	--
	Bedroom #2	Ceiling light	--	75W	--	--	--
		Conv. rec.					
		3 @ 180W	--	540W	--	--	--
3	Living room	Air cond.	--	--	1400W	--	--
	Bedroom #1	Air cond.	--	--	1200W	--	--
	Bedroom	Wall heater	--	--	1400W	--	--
4	Kitchen	Refrigerator	--	--	--	300W	--
		Toaster	--	--	--	1000W	--
		Iron	--	--	--	1000W	--
		Totals					

A Typical Circuit Survey
Figure 8-5

lamps. For convenience receptacles with no special purpose, assume 180 watts as the average load. If exact wattage is needed for equipment, a wattmeter can be used.

Your completed survey form is like a doctor's X-ray. It gives a good picture of the electrical system. It's an excellent way to diagnose problems and prescribe remedies.

Walk Before You Run
I would be omitting a major item in discussing remodeling work if I didn't issue one important caution before going on to the next subject. Remodeling is different. It should carry higher markups because there's more risk involved. After a few jobs you'll learn to protect yourself from the most common mistakes electrical contractors make when working remodeling and rehab projects.

But those first few jobs can be very expensive training. You'll minimize the chance of a major loss if you begin with smaller jobs. Select them carefully. Try to find remodeling work that's similar to the new construction you've been handling and can be completed in a relatively short time.

When you've accumulated experience on several smaller, simpler jobs, take on more complex work. You'll have the experience and confidence needed to do a professional job and earn a professional's reward.

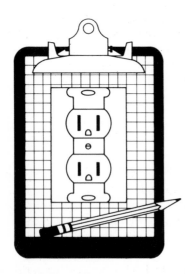

Chapter 9:

Three-Phase Wiring

Single-phase alternating current wiring is commonly used for smaller system loads. For example, a home with 100 amp service would use single-phase circuitry. But most power is generated and distributed nearly to the point of use in three phase circuits by the utility.

Industry, farms and commercial buildings with large loads also use three-phase wiring because it's more efficient than single-phase. Three-phase conductors weigh less than single-phase conductors of the same power rating. Three-phase motors cost less, weigh less and are more efficient. It's also simpler to balance large loads with a three-phase system.

Differences Between Single-phase and Three-phase
The basic difference between single-phase and three-phase circuits is in the number of alternating currents. In a three-phase circuit, there are three. In a single-phase circuit there is only one. The difference in the number of currents depends on the generator.

A three-phase generator is built with three separate windings. These windings are spaced so that each one occupies 120 degrees of the rotor, or one-third of its circumference. When the rotor turns, its windings cut through the magnetic flux line produced by the field poles of the generator, producing voltage in each of the three windings.

Because the windings are displaced 120 degrees from each other, the voltage produced in any one winding will be 120 degrees ahead of the voltage produced by the next winding. It will be 120 degrees behind that in the third coil. Therefore the three voltages produced are spoken of as being 120 degrees out of phase with each other.

For planning and design purposes, the three windings are labeled *A, B* and *C. A* is assumed to be the lead voltage, followed by B and C. Each of the three windings carries the design voltage of the generator.

Types of Connections
There are two kinds of three-phase connections. One is called the *wye* connection. You'll see this written as a *Y*. The second is called the *delta* connection and will usually appear as a triangle on the plans. Generators, transformers and other loads may be connected either in wye or delta. Advantages of the two connections will be discussed after their differences are explained.

The Wye Connection
The word *wye* is taken from the layout of this connection. Vectors which represent the voltage in each winding are drawn on paper 120 degrees apart. All three vectors originate from a common point. The vectors form the letter *Y* when drawn in schematic form. See Figure 9-1.

In Figure 9-1 the circle represents the power source. The three lines coming out of the circle to the right connect to the loads you plan to install. The coiled line represents the coils of the generator or transformer. Notice that these coils originate from a common point and end at the connection with the distribution wiring. This distinguishes wye and delta type systems, as will be explained later.

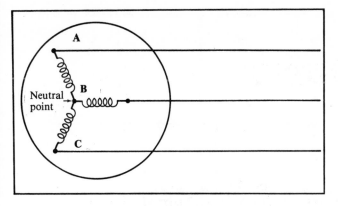

A Wye Schematic
Figure 9-1

A conductor is connected to each of the winding ends at the generator or transformer. The other ends of the windings are joined together, as noted before, at a common point. This common point is called the neutral point. In a four-wire, three-phase system, a neutral wire comes off the neutral point. In a three-wire, three-phase system, no wire comes off the neutral. Either three-wire or four-wire circuits can be used, depending on the application.

When the voltage windings from a source—either a generator or transformer—are connected in a wye, two levels of voltage are available for use. One level is the voltage generated in each phase winding. This is the voltage between the phase conductor and the neutral point. The other level is the voltage between any pair of the three conductors.

Compute the voltage in a wye system using the following:

Phase-to-neutral voltage = source rating

Phase-to-phase voltage = $\dfrac{\text{phase-to-neutral voltage} \times 1.73}{}$

(1.73 is a mathematical constant)

For example, if a wye-connected generator produces 4,000 volts in each phase winding, that much voltage would be available from any phase conductor and the neutral point. The voltage between any two phase conductors would be 4,000 times 1.73, or 6,920 volts. This type of generator would be labeled a 4,000/6,920 volt, three-phase 3 wire generator.

Remember that two voltages are available for *each* of the three phases. For the above example,

there are three phases of 4,000 volts and three phases of 6,920 volts.

The current in any one phase conductor from the source to the load is equal to the current flowing through the corresponding phase winding in the source. This is stated by the formula:

$$I_1 = I_p$$

I_1 = the line current and I_p is the phase winding current. When the load is connected to the conductor, the current flows from the phase winding through the conductor to the load and then back to the winding. This current flow takes place with or without a neutral conductor.

Whether a neutral wire is used depends on the voltage needed. When the entire connected load is rated at the line-to-line voltage and all loads are connected to all phases, no neutral wire is used. For example, when three-phase motors or transformers with the same voltage as the power source are the only load, no connection to the neutral point is needed. Figure 9-2 illustrates what is called a balanced load.

But suppose single-phase loads must be connected to that three-phase circuit. See Figure 9-3. Three loads in the drawing use 120 volts, which is 208 divided by 1.73. Single-phase loads such as appliances, light fixtures and single-phase motors operate at various loads and don't all run at the same time. That would unbalance the system, placing a relatively heavy load on one phase and a relatively light load on others. The neutral wire carries excess current back to the power source and balances the load between the phases. Note the arrows indicating excess current returning to the transformer at the lower right of Figure 9-3.

The phase-to-neutral voltage in this type of system is usually 120. That's a suitable voltage for many single-phase loads. This type of system is called a 120/208 volt, four-wire Y system. It's used in many large apartments and commercial buildings.

A balanced Y system that doesn't need a neutral wire for voltage changes may still use the fourth wire. The ground to the neutral point provides an important advantage: A short in any of the phase conductors will open the breaker before a phase-to-phase fault develops. This method can also be used to supply a single-phase load which is rated for the phase-to-phase voltage of the Y system.

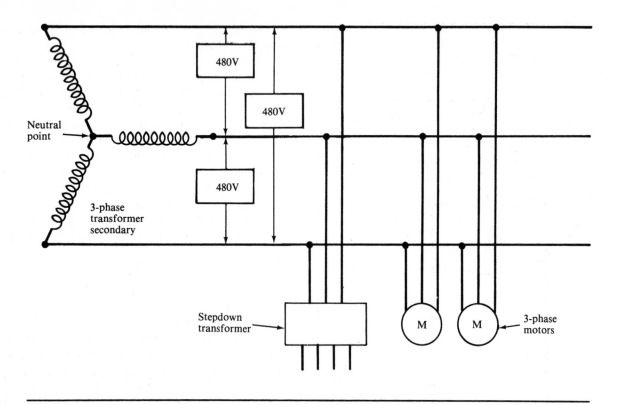

A Three-Wire Wye System With A Balanced Load
Figure 9-2

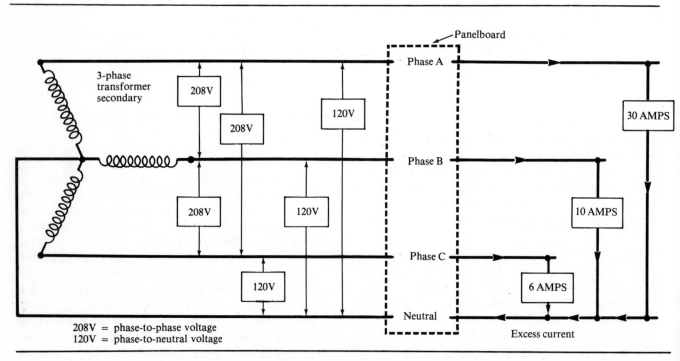

208V = phase-to-phase voltage
120V = phase-to-neutral voltage

A Four-Wire Wye System
Figure 9-3

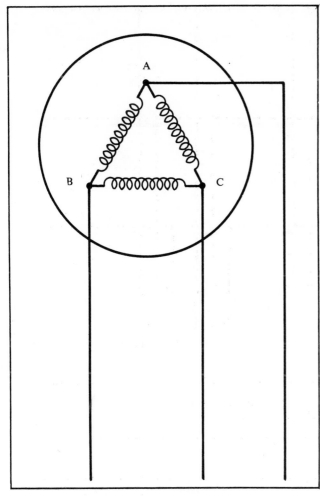

A Delta Connection Schematic
Figure 9-4

E_1 is the line voltage and E_p is the voltage in the phase winding.

The current relationships in a delta system are like the voltage relationships in a Y system. The current flowing in each phase winding of the delta system has one value called the phase current. But each line conductor going to the load is connected at a junction of two of the phase windings. As a result, the current flowing in the line conductor to the load is the average of the two phase currents.

With a balanced load, the phase current will be the same in all three windings. The line current will also be the same in all three lines. The relationship between phase and line current for a balanced delta system is:

$$I_1 = I_p \times 1.73$$

I_1 is the line current and I_p is the phase current.

As a rule, delta sources are used only for three-phase, three-wire loads. But a delta system can supply both single- and three-phase loads if a fourth conductor is connected at a mid-point in one of the windings. See Figure 9-5. This is usually called a "red leg" delta system. The typical voltage in this system is 240 volts from phase-to-phase and 120 volts from the mid-phase neutral conductor to each of the phase conductors.

The Delta Connection
In the delta connection, the three 120 degrees displaced windings of the generator or transformer are connected end-to-end. See Figure 9-4. This forms a triangle and that's the symbol you'll see on plans and diagrams. The name comes from the Greek alphabet letter *delta* which looks like a triangle.

In a delta circuit, a conductor is tapped from each winding junction. The windings are continuous within the power source. Only one voltage is available from each phase winding. This phase-to-phase voltage is often spoken of as the "line voltage," since its value is the same as the voltage between each pair of line conductors. This is expressed by the formula:

$$E_1 = E_p$$

Types of Service
The system designer must decide whether to use single or three-phase service. It costs more to wire a building for three-phase power. Service costs may also be increased, since it generally costs the utility more to furnish three-phase current. But a three-phase motor of a given size usually costs much less than a single-phase motor of the same size. If power cost is a major consideration, using three-phase motors will save money in the long run.

In any case, the utility will be able to advise you on the most economical choice. Some power companies may require that the service be three-phase. Contact the utility early in the planning stages to avoid wasted time and effort.

If three-phase service is needed, the next question is whether it will be a delta or Y system. This answer depends on what the utility is willing to provide and the types of loads to be serviced.

Where single-phase loads make up the bulk of the load to be serviced, a four-wire system is best. The phase-to-neutral voltages supply common single-phase loads such as lights, appliances and small motors. Three-phase circuits are available for three phase motors and other devices.

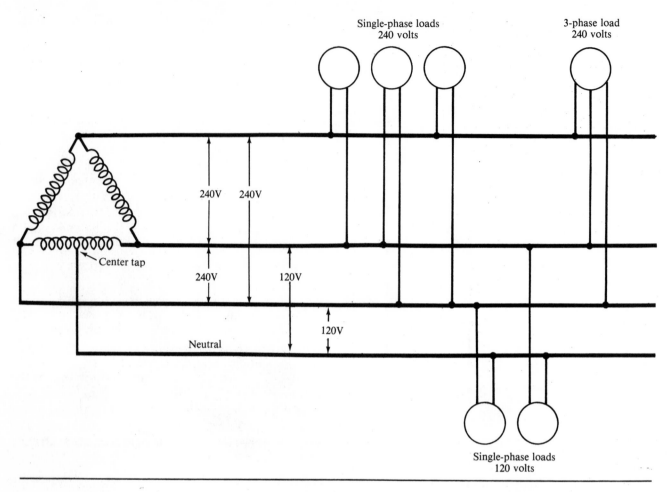

A Delta Connection Tapped To Serve Single-Phase Loads
Figure 9-5

A common three-phase, four-wire Y system is the type rated at 120/208 volts. With this system, it's important that the three-phase motors are actually rated at 208 volts. Don't use any other voltage motor. For example, a 220 volt motor in a 208 volt circuit would have an undervoltage of 5.5 percent of the running current. But starting current draw would be much larger. Circuit overloads would be a constant problem.

You might be able to use the 220 volt motor on a 208 volt circuit by minimizing voltage drop. Either reduce the distance between the panel and the load or increase the wire size.

Electric ranges and other cooking units in a 120/208 volt system should have 208 volt elements. If not, install an autotransformer to furnish 120/240 volt service from a 120 volt phase of the circuit.

As mentioned earlier, delta sources are usually used only with three-phase, three-wire loads.

However, they can also serve single-phase loads with a neutral conductor tapped at the mid-point of a winding. Never create a circuit that connects the neutral conductor and the third or "wild" phase conductor. This is the conductor that connects at the ends of the two untapped windings. It will have a higher voltage than the other phase conductors. The NEC requires that this conductor either be orange or be tagged if the neutral is also present.

Delta systems have the advantage of furnishing the standard voltages of 120 and 240. However, they tend to unbalance three-phase service and cause circulating currents in the transformer if the single-phase load is large compared to the three-phase power available.

Design and Installation
Installation procedures are the same for both three-phase and single-phase wiring. The same con-

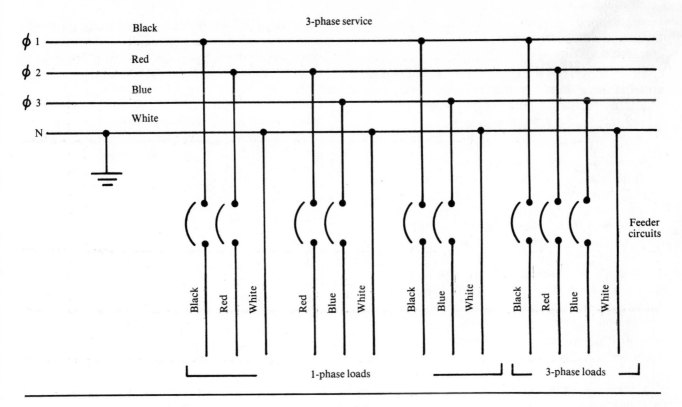

Balancing Single-Phase Feeders
Figure 9-6

siderations apply. Meter and service equipment is different, so be sure you order the type that is both appropriate and approved by the power company.

Feeder and service loads for both single-phase and three-phase wiring are calculated as described in Chapter Six. NEC Section 220-B has special requirements when installing two or more single-phase ranges supplied by a three-phase, four-wire feeder. This section allows the total load to be calculated on the basis of twice the maximum number connected between any two phases. See Example 7 in Chapter 9 of the NEC.

Phase Converters
Phase converters let you run three-phase induction motors off a single-phase power source. Some utilities limit the size of single-phase motors that converters can serve. These limits may also apply to three-phase motors supplied through converters. Ask the utility before installing a phase converter.

Phase converters present no unusual problems for steady loads with low starting torque. Such loads are typically large fans and turbine-type pumps. Equipment that needs a high starting torque or is subject to wide fluctuations needs special consideration. Get good advice from the manufac-

turer of the motor and the converter. Compressors and pumps that start under load need high starting torque.

The starting current of a combination converter and three-phase motor is generally less than that used for a comparable three-phase motor alone. The starting torque is also lower. But sometimes you'll have trouble providing overcurrent protection because of the longer starting period and the unbalanced current in the motor windings. Consult local codes and the manufacturer. You may be able to design the system so that only the overcurrent devices that will do the job are built into the converter at the manufacturing stage. This will reduce the cost.

There are two general types of converters, the static converter and the rotary. It's only important that you match the converter to the motor. Follow the manufacturer's recommendations. A rotary converter needs a horsepower rating as large as that of the largest motor to be driven. Other smaller motors can be supplied by the same converter.

Rotary converters must be running before starting any motor. Motors are then started and stopped independently of each other. A minimum motor load of one-tenth the converter rating is recommended. A total motor load of about two times the converter rating in horsepower can be connected.

Static converters are often used for single motor applications. They're usually matched to the rating of the motor. That is, a 5-horsepower converter would serve a 5-horsepower motor.

Color Coding and Balancing Loads

A three-phase system used to supply a number of single-phase loads will use four wires—the three-phase wires and a neutral. Single-phase systems are then tapped off the three-phase system. For example, in a large apartment building, the service into the building may be three-phase, four-wire. Once inside it is broken down into single-phase feeders to each apartment. These feeders will use two of the phase wires and the neutral for a three-wire feeder service.

Your objective is to try for reasonably balanced loads on the three-phase service. Where possible, take the same number of taps off each of the three-phase wires as shown in Figure 9-6. There's no way to avoid some unbalanced current flowing in the neutral. But the operating efficiency of the system should be satisfactory.

Avoid confusion and mistakes in connecting wires by color-coding them. The colored feeder wire would be tapped to the same color in the three-phase service. When connections are made at a common point, assign the phase wire number the feeder wire color. Devise any color system that meets your needs and makes sense to you. But follow good wiring practice and the NEC. Another good idea is to draw up a feeder panel schedule and then refer to it as the job progresses. That reduces the chance of an error.

Lighting Loads

For heavy lighting loads, a 277 volt system may be the best choice. This system will usually use a 480 volt three-phase line in a four-wire Y connection. Phase voltages of 277 volts are then available between any phase wire and the neutral.

Restrictions on these high voltage lighting systems are outlined in NEC Section 210-A. Generally, they're permitted only in industrial locations, stores, office buildings and the like. The fixtures and wiring must be approved for high voltage installations. Incandescent fixtures have to be mounted at least eight feet above the floor. Conditions on site must indicate that only qualified persons will service the fixtures.

Chapter 10:

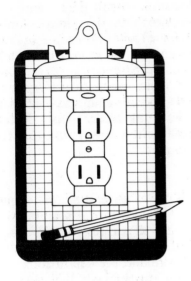

Starting Your Business

The chapters you have read so far deal primarily with how to do the electrical work. But this isn't just a guide for electricians. It's a manual for electrical contractors. And you need to be more than just a good electrician to make it in the electrical contracting business.

That's why this chapter, and most of the rest of the book, explains how to make a good living as an electrical contractor.

Getting Started

Starting a business is usually much easier than keeping it going. Each year new businesses of all types are started. Each year many more of them fail than succeed.

Usually it's hard to identify a single or even a few reasons why a contracting business succeeds. It's easier to see why a business fails. Probably the two most common reasons electrical contractors fail are poor record keeping and lack of operating capital (cash). Other reasons include a slump in building activity, lack of business experience, and poor bidding practice. Notice that this list doesn't include lack of skill or technical know-how. Your business will succeed or fail on your ability as a planner and manager, not on your skill as an electrician.

Running an electrical contracting business requires huge financial risk, a major commitment of time and effort, and considerable personal sacrifice. Yet many of those that fail have approached their business venture as casually as if they were planning a two-week vacation. Make a

commitment now to do a good job of planning your entry into the electrical contracting business.

Start with your ambition, hopes and dreams. But create a realistic, hardheaded, practical plan before going into business for yourself.

At some point as you go through this chapter you may begin to feel that all this planning and all these details are just too much work. What you want to do is just start taking on jobs and making money. That's the best indication you'll get that electrical contracting may not be for you. Running a contracting business takes planning and attention to detail—all the time.

Not everyone is cut out for running a business. And if you're not one that is, don't worry about it. Electricians make good money. And many electrical contractors would actually make more money if they worked for wages on someone else's payroll.

If you're about to start an electrical contracting business, read this chapter carefully and follow the recommendations given here. This chapter should give you a good grasp of key procedures. As your business grows, sound planning will become a habit and serve to protect you and the strength of your company. Of course, there's no guarantee of success for any business, but good planning will tip the odds in your favor.

Your Business Objective

When you announce to the world that you intend to start your own electrical contracting business, someone will ask why you're doing it. On the surface the answer seems obvious. Yet it's surprising the way that question gets answered:

"I know what my boss does. If he can make money at it, I know I can."

"My Uncle Jim has his own business and he spends every winter in Florida. It seems like the thing to do."

"I've worked for somebody for 15 years. I know I can make it on my own."

I've heard these reasons given by potential new electrical contractors. And while they're honest responses, they don't inspire much confidence in the person who asked the question. Start your contracting firm with a more realistic appraisal of what you intend to do:
• You should be able to explain your reason in one or two sentences.
• It should be positive, not negative.
• It should reflect your understanding of existing opportunities.

Just for a moment, imagine that you're general contractor or the owner of a new building that will need wiring in a few months. Imagine that you're talking to a new electrical contractor who's explaining why he's getting into the business. He says, "Well, I've been working for the same outfit for ten years. I do good work but I still get laid off in winter. I might as well go to work year 'round for myself."

Compare that answer with this one: "There's a lot of work available now that the big boys are passing up and some of the smaller firms don't have the skills to handle. I know how to handle these jobs and can get a good share of them if I keep my overhead down."

Which answer would impress you if you were a general contractor or owner? Either would certainly consider other information before taking a chance on your firm. But the point here is that the man with the second answer probably has a proposal that's worth listening to.

There's nothing wrong with taking pride in your ability. And ambition shows a willingness to work hard to achieve your goal.

Draw up the reason you are going into contracting. But don't go overboard. Keep your reason simple, forceful and realistic. Above all, be sure it is *your* reason. Don't try and fool anyone, least of all yourself.

But recognize that there's only so much work around. If the electrical subcontractors presently active in your area can't handle it and you can, that's reason enough to join their ranks. If there isn't enough work to keep every electrical sub busy, you're going to have to find some way of doing the work better, cheaper, faster or smarter to take jobs away from the competition.

Decide what your area of interest is and realistically appraise your chances of either handling the overflow or taking jobs from others. The next section explains how to appraise your chances of success.

Market Study
Now that you have a reason for going into business, let's see if your business venture is practical. You'll be able to control your business most of the time, but success also depends on factors outside your control. Take a hard look at these before going any further. All the capital, hard work and personal sacrifice won't make up for a bad market for your work.

A market study evaluates the chances of finding enough buyers for a particular product or service to make selling that product or service worth the effort. Market studies are very complicated and costly for larger companies that are experimenting with a new product like beer or toothpaste. Your study can't be that complicated, and it doesn't have to be. It can be inexpensive, simple and yet accurate.

But whatever the size and no matter how it's done, record the results in a neat, logical manner. It becomes the foundation of your business plan and will be reviewed by people who are important to the success of your venture.

Start your market appraisal by visiting the county clerk or building department where building permits are recorded for your area. It's a good idea to call before you go the office so you know the rules established for using the records on file. But these are public records that must be available to everyone.

Go back through the permits for several years and try to determine if the value of construction is increasing or decreasing in the area where you plan to do business. You don't need to examine every project, and you may want to skip certain months. Just look at permits for the type of work you plan to handle. Go back far enough to establish a trend over both good and bad economic times.

Is the monthly total of permit values increasing or decreasing for the work you plan to handle? How does it compare with building activity nationwide? Is one type of work becoming more common while others are becoming fairly rare?

You'll get a pretty good idea of how business is going for electrical contractors and where the growth is when you've spent several hours going over old permits. Certainly you'll know more than could be learned by watching for construction projects as you drive through the area. And you'll have more hard facts about construction volume than many who have been active in electrical contracting in the area for many years.

Next, count the electrical contractors listed in the yellow pages of your phone book. Find an old phone book for the area and make the same count. Many libraries have good collections of phone books, including older editions.

If you plan to do work in an incorporated city, the city hall will probably have a list of companies licensed to do business in each of the last several years. This is an especially valuable record for your survey. It should show the name of every electrical contractor who took out a permit in that city for the last several years and the dollar volume of work done in that city for each year. The list will include many contractors who aren't in the phone directory but who came into the area to do work. Again, count the number of electrical contractors listed. Is the number increasing or decreasing? Is the volume of work increasing or decreasing?

Here are some other good sources that will indicate the prospects for electrical contractors in your area:

• Nearly all major projects go through a lengthy permit process in most states. The local planning department can tell you what major jobs are in the permit process and where major subdivisions are planned.

• Applications for sewer, electric, gas, and water permits indicate future activity. Check with the department or utility for planned extensions of service.

• Are any major employers planning new facilities in your area? The new plant will attract employees. Where will they live?

• If you live in the city, be alert to building permits issued for remodeling in older neighborhoods. That means owners are investing in their properties and values are probably increasing—a good sign that there will be work for electrical contractors in the area.

• What population growth can you expect in your area? Your local Chamber of Commerce office can probably supply details. Many real estate agents prepare market surveys to help sell major pieces of property. These usually cover both population trends and economic prospects. The survey is usually discarded after the property is sold. Ask several agents if they have information on population growth and employment in your area.

• The U.S. Census Bureau is probably the best source of statistical data for market surveys. Besides national data, the Census Bureau has *Census Tracts* and *Standard Metropolitan Statistical Area (SMSA)* designations available. These studies divide large cities and adjacent areas into tracts of about 5,000 residents for the purpose of showing different statistics. Census Bureau data includes a wealth of information on income and housing in each census tract.

The Bureau compiles the number of permits issued in approximately 5,000 communities and publishes this data monthly in "Housing Units Authorized by Building Permit and Public Contract." A good public library will have back issues for your use. A yearly subscription is $50 from the Superintendent of Documents, U.S. Government Printing Office, Washington, D.C. 20402.

Every 10 years the Bureau conducts a census of housing. Information is compiled on the year each residence was built, the number of rooms, owner cost, value, monthly rent, condition and type of plumbing. This data is available for about 300 larger cities and for each of the thousands of census tracts. Your state census data center will be able to supply either computer printouts or printed reports for the census tracts you plan to cover. The U.S. Bureau of the Census can supply the phone number of the data center in your state.

Public Information Office
Bureau of the Census
Dept. of Commerce
Washington, DC 20233
(301) 763-4040

The Office of the Census of Housing at (301) 763-2873 and the Construction Statistics office at (301) 763-5731 can answer your questions about what information is available.

When you've investigated all the data that seems relevant, prepare your market study. Figure 10-1 is an example. Add a brief written summary and attach it to the study as an appendix. Include quotes from any authoritative source that help define the market for electrical contracting services in your area.

For example, after the information shown in Figure 10-1, you may want to add:

"John O'Malley has been a residential builder the past 20 years here in Beaver City. It is Mr. O'Malley's assessment that the increase shown in

	1980	**1981**	**1982**
January	8	12	12
February	5	10	14
March	7	14	18
April	10	18	19
May	14	24	31
June	16	28	39
July	20	33	42
August	22	31	43
September	19	30	40
October	21	29	33
November	15	22	25
December	11	18	22
Annual Total	168	269	338

Electrical Contractors Beaver City Yellow Pages	1980	1981	1982
	8	8	8

Average number of homes wired per contractor	1980	1981	1982
	21	33	37

Number of subdivisions in Beaver City area with preliminary approval:

4

Total # of lots:

62 - 30 residential
32 duplex & condominiums

Market Study
Figure 10-1

home construction will continue for at least another five years. This growth rate could, in Mr. O'Malley's estimation, easily accommodate another electrical contracting business."

Be sure the quote is accurate and get permission from the person quoted to use the words in quotation marks.

Estimating Gross Sales

If you have determined that there's a need for your electrical contracting business, the next step is to estimate the minimum annual sales you'll need to operate. This information is very useful. It will serve as a guide for your daily operations and is your *estimated income statement* when you go to the bank for financing.

A quick way to estimate annual sales is to begin with your expected gross profit figure. For example, in the construction industry, gross profit is usually about 10%. Let's say you want to make

$10,000 gross profit in your first year of operation. This is over and above any salary you pay yourself. Divide $10,000 by 10%:

$$\frac{\$10,000}{.10} = \$100,000 \text{ annual gross sales needed}$$

$$\frac{\$100,000}{12} = \$8,333 \text{ monthly gross sales needed}$$

While this is a quick way to estimate the sales you'll need, it isn't very helpful in running your company. Rather than use the expected profit as the base for your calculation, draw up your estimated annual operating expenses as in Figure 10-2. Most of these expenses can be estimated fairly accurately. It will take five times the total amount of these expenses in sales to stay in business, if the markup on your material and labor expense is 10%.

Item	Amount
Owner's Salary	$21,000
Bookkeeper	5,000
Rent and Heat	4,800
Electric Bill	720
Depreciation on Equipment	2,500
Interest on Loan	6,000
Telephone	600
Truck and Travel Expenses	5,000
Insurance	1,500
Office Supplies and Miscellaneous	500
Contingency Fund	1,000
Total	**$48,620**

Estimated operating costs per year — Sparks Electric Company

Typical Operating Costs
Figure 10-2

Annual Operating Cost	$48,620
	x 5
Annual Sales Needed	$243,100

If you have to borrow money to get started, your estimated income statement and market study are essential. They show the loan officer that you know how much revenue is likely and how much is needed to stay in business.

The figures on both these forms must be reasonable and based on fact. No banker will take your word that you're not guessing. He has statistics which list average income for all sizes of electrical contracting firms in all parts of the country. It takes only a moment to check your estimate against these. He is probably well aware of the local business climate. He deals with people in the area every day. His or her experience will help confirm that your figures represent an accurate picture.

But the loan officer is really trying to determine if he has confidence in you. If your figures can't be believed, how can he accept your promise to repay the loan? If the market study and income statement are amateurish and based only on dreams and hopes, you probably don't have the business sense to make money in the electrical contracting business.

Starting Costs
Besides operating costs, you always have starting costs when you open a new business. Starting costs

are typically one-time expenses that take place when you begin operation. Figure 10-3 shows what the minimum costs might be for a new electrical contracting firm in 1983 dollars. The figures will be much higher as inflation erodes the value of the dollar.

Figure 10-3 assumes you rent a small office and warehouse, but doesn't include any receivables. Remember, there may be several months between the time you get your first good job and the time you get paid for the work done. It will take 30 to 60 days to collect on most jobs. Receivables may average about 15% of your annual gross receipts. If you expect to take in $200,000 your first year, expect money owed you to average about $30,000. Naturally, you have to pay your employees and have money to live on while you're waiting for payment. Don't underestimate the importance of this burden.

Personal Balance Sheet
Besides the market feasibility, starting costs and operating costs, any loan officer will want to know something about your personal worth. He'll ask for a personal financial statement or balance sheet. Most banks ask that you use their form. It will look something like Figure 10-4.

The balance sheet has two sections:
• Assets - things you own
• Liabilities - things you owe

As the name suggests, the assets and liabilities balance (are the same amount) when all figures are

Item	Amount
Inventory	$40,000
Tools & Equipment	5,000
Office Equipment & Fixtures	1,500
Legal Fees	200
Accountant Fees	100
Utility Deposit	150
Building Security Deposit	1,000
Insurance	500
Miscellaneous	100
Contingency Fund	2,000
State/Federal Employer Deposits	4,000
Truck (Used)	3,000
Total	**$57,550**

Estimated starting costs — Sparks Electric Company

Typical Starting Costs
Figure 10-3

added. Figure 10-5 is a condensed balance sheet. Obviously, no one owes exactly what he owns. The difference between your assets and liabilities is your *net worth*. Net worth is listed as a separate entry in the liability column of the balance the sheet.

Besides the assets and liabilities, the loan officer will want an income and expense statement, showing all sources of income and where the money was spent. He'll also want a list of all other loans outstanding and loans you have paid off in the last several years.

Be as thorough as possible in preparing your financial statements. Assets should include anything of value which you own—cash savings, stocks and bonds, real estate, vehicles and the like. Liabilities should include any formal debt or any debt over $50. This includes credit cards and any personal loans made on a casual basis. Remember that lenders like to be thorough. Any omissions, even if they are honest mistakes, make the loan officer look harder for other mistakes. Too many of them will disqualify you from consideration.

How Much Money?
The question of how much money to ask for is one with several answers.

As the business owner, your inclination will be to borrow as much money as possible. In your mind, getting the largest loan possible saves you from investing your last dime in the business. It leaves some of your assets intact for future use. Although it may not seem important at the time, a private "nest egg" of $3,000 or $4,000 may be a life saver at a later date.

Here's another reason to borrow a larger amount. Nearly every estimate of expenses omits some important item. You'll probably need more money than your estimates indicate. Having enough cash on hand will save you the embarrassment of another trip to the bank.

Your banker, on the other hand, will also have a figure in mind. It will probably be less than yours, for several reasons. First, he wants you to invest a sizable portion of your own assets in the business. This reduces the bank's risk. It also guarantees that you'll do everything humanly possible to make the business succeed. You would suffer a good deal if it fails. Second, his figure will be limited by interest rates, the amount of money the bank has to loan and the assets you have to secure the loan. And finally, you'll get only the amount of money that can be paid back from your monthly sales volume.

One way to calculate a minimum loan figure is to take your starting costs and add six months of operating costs. Referring to Figures 10-2 and 10-3, for example, a minimum loan for James Sparks to start his business would be:

Starting Costs	$57,550
Operating Costs - 6 Months	24,310
Total	$81,860
Less Personal Equity	- 5,000
Total Loan Needed	$76,860

CONFIDENTIAL FINANCIAL STATEMENT OF

TO

NAME | DATE OF STATEMENT

HOME ADDRESS

BUSINESS AFFILIATION

(NAME OF CONCERN) | (ADDRESS)

IMPORTANT: ALL QUESTIONS MUST BE ANSWERED. IF INAPPLICABLE WRITE "NONE"

ASSETS					LIABILITIES			
Cash on Hand and in Bank	$			Unsecured Notes Payable	This Bank	$		
Notes Receivable—Good—Unpledged					Other Banks (NAME)			
Accounts Receivable—Good					Individuals			
Listed Stocks—Bonds—Schedule B					Other			
Other Current Assets (List Separately)				Collateral Loans—Secured by Stocks, Bonds, etc-Sched B				
				Accounts Payable				
				Mortgages Payable—This year—Schedule C				
				Other Lien or Installment Debts				
				Taxes Unpaid—Schedule D				
				Other Current Liabilities (List Separately)				
TOTAL CURRENT ASSETS				TOTAL CURRENT LIABILITIES				
Cash Surrender Value—Life Insurance—Schedule A				Mortgages Payable—After this year—Schedule C				
Unlisted Stocks—Bonds—Schedule B				Chattel Mortgages				
Real Estate—Homestead—Schedule C				Land Contracts				
Other Real Estate—Schedule C				Other Liabilities (List Separately)				
Other Assets (List Separately)								
				TOTAL LIABILITIES				
				NET WORTH				
TOTAL $				TOTAL $				

CONTINGENT LIABILITIES: Notes Receivable discounted with your endorsement (not included above) ... $

Accommodation Endorser or Guarantor for others (describe) ... $

Other Contingent Liabilities, Unsatisfied Judgments, Pending Suits (describe) ... $

Is your broker carrying Stocks for you? ... If so describe

BANKING CONNECTIONS ... Loan

$ ... Secured by:

$ ... Secured by:

SOURCES OF ANNUAL INCOME 19___			ANNUAL OBLIGATIONS 19___		
Salary or Compensation			Mortgage Principal		
Bonuses			Mortgage Interest		
Dividends			Property Maintenance		
Other income (Alimony, child support or separate			Property Taxes		
maintenance income need not be revealed if you			Income Taxes		
do not wish to have it considered in determining			Premium on Life Insurance		
your credit worthiness).			Rent		
			Installment Purchases		
			All other Loans—Principal		
			All other Loans—Interest		
			Other Obligations (specify)		
TOTAL ANNUAL INCOME			TOTAL ANNUAL OBLIGATIONS		

PERSONAL (INDIVIDUAL) FORM 1900 (9/77) H.C. MILLER CO., MILWAUKEE   (OVER)   Page 1

Bank Personal Credit Form
Figure 10-4

PLEDGED ASSETS—If any of the assets below have been pledged, place an (X) before the item.

SCHEDULE A: Life Insurance Carried:–

Pledged	NAME OF COMPANY	BENEFICIARY	FACE OF POLICY	CASH SURRENDER VALUE	LOAN	

SCHEDULE B: Listed Stocks, Bonds, Etc., Owned:–

Pledged	DESCRIPTION	COST	MARKET
	TOTAL		

SCHEDULE B: Unlisted Stocks, Bonds, Mortgages, Etc., Owned:–

Pledged	DESCRIPTION	COST	MARKET
	TOTAL		

SCHEDULE C: Real Estate Owned and Mortgages Payable:–

Pledged	TYPE, LOCATION AND DESCRIPTION	TITLE IN WHOSE NAME IF JOINT SPECIFY	VALUE			MORTGAGES		INSURANCE		RENTS REC'D DURING YEAR
			ASSESSED	APPRAISED	FORCED SALE	AMOUNT	DUE	FIRE	TORNADO	

SCHEDULE D: Taxes Unpaid:–

NATURE OF TAXES	FOR YEAR	WHEN DUE	AMOUNT

STATE OF WISCONSIN

COUNTY OF _____

    Being duly sworn, the undersigned declares that the above information is true, correct and complete as of the date of this statement and is provided to the Bank for the express purpose of inducing it to extend present or future credit to the undersigned, the undersigned and another, or to another guaranteed or endorsed by the undersigned. The Bank, or its agents, may verify this information and obtain additional information concerning the undersigned's credit standing and furnish the same to others. This statement is the Bank's property. The undersigned agrees to notify the Bank promptly of any material change in this information.

_____ (SEAL)

_____ (SEAL)

_____ (SEAL)

    Subscribed and sworn to before me on the date of this statement.

_____

Notary Public, _____ _____County

My commission expires _____, 19 _____

Bank Personal Credit Form
Figure 10-4 (continued)

Personal Balance Sheet 1/15/83

James Sparks, Electrician

Wattville, Ohio

Assets:

Cash on Hand	$300
Cash in Bank	3,000
Accounts Receivable	1,500
Securities	2,000
Real Estate	55,000
Automobile	5,000
Other Assets:	
Tools and Equipment	3,000
Personal Chattel	12,000
Life Insurance	500
Total Assets	**$82,300**

Liabilities:

Accounts Payable	-0-
Loans Payable	8,000
Real Estate Mortgage	21,000
Real Estate Taxes	3,000
Interest Payable	600
Other Liabilities	-0-
Total Liabilities	**$32,600**
Net Worth	**$49,700**
Total Liabilities and Net Worth	**$82,300**

Personal Balance Sheet—Page 1
Figure 10-5

Keep in mind that this figure may vary a good deal. Your banker may offer you less than this to get you to invest more of your money (if you have it). On the other hand, if you have a good source of collateral—a home that's paid up, for example—you may want to ask for more money. It should be obvious if you are in a position to bargain. If so, don't be afraid to ask for more. If you are refused, ask for reasons. Counter them with your own. Offer reasons why you think you should have more money. Finally, don't forget that it's perfectly legitimate, even prudent, to shop for a loan like you would shop for a new car.

In recent years banks have become much more competitive. *If you have a sound financial posi-tion*, there's no reason why you have to take the first offer.

Finally, a word of caution. Whatever the amount of money you get, use it wisely. Don't, in your first burst of enthusiasm, make a lot of purchases that you'll later regret. Follow your cost estimates as close as you can. Before buying anything, shop around to get the best price and best terms of payment. This is especially true when building your inventory. Some materials suppliers offer extended payment terms to build customer loyalty if you are a good risk.

Money Definitions

Like any other tool, money can be put to work in

Sparks - Balance Sheet - Page 2

Banking:
Checking Account #322-010	$ 300	Wattville State Bank
Savings Account #679-034	$3,000	Wattville State Bank

Life Insurance:
Policy #384-7-766-1	$ 500	Ohio Mutual Insurance Co.

Securities:
200 shares	$10 Book Value	Allied Wire & Supply Co.

Accounts Receivable:
$1,500 due 3/30/82	Gilmore Const. - wiring work	Unsecured

Real Estate:
Single-family dwelling	Assessed value $55,000	1100 Main St., Wattville, Ohio
1st mortgage-20 year	$21,000 due	Buckeye Savings & Loan, Wattville, Ohio

Loans Payable:
$8,000 + $600 interest	Payable at $180 monthly until 1987. Home improvements & college tuition for children.	John Sparks, father, lender. Pittston, Ohio

Taxes
$3,000 due quarterly in 1983	Property tax	City of Wattville

Personal Balance Sheet—Page 2
Figure 10-5 (continued)

many ways. The following terms are used to define how cash is being used.

Permanent Capital

Permanent capital is money which doesn't have to be paid back. Instead, it earns you, the owner of the business, the right to share in the profits. For example, you invest $10,000 of your own money in your business. It becomes a part of the business. In return you receive any profit made at the end of the year.

Permanent capital can also be called *equity capital*, or simply equity. Equity capital is a permanent part of the business, not a loan. Equity can be increased by putting more of your or someone else's money in the business.

Banks never lend equity capital—or at least they try not to. Some small business investment corporations are willing to invest equity capital. Equity can also be generated by selling off inventory or unwanted equipment.

Working Capital

Working capital is the lifeblood of your business. It is short-term money that bridges the gap between expenses and payments. Few small businesses make money before it's needed. Until that money comes in, you need cash to pay bills and meet payroll.

Suppose you bid and get the contract to wire a 48 unit apartment complex. This means hiring more men and increasing inventory. But you don't get even a partial payment until the job is 50% completed. Until then, your working capital has to carry the company.

Unlike equity, short-term capital can be in the form of a loan. Interest is charged at a higher rate than for long-term loans. Payback may be one lump sum with interest.

Where construction work is seasonal, contractors may apply for working capital loans in spring, after contracts have been let. If you deal with the the same bank over a period of years, getting a small short-term loan may take only a phone call and few minutes at the bank to sign the papers.

Sometimes you can avoid borrowing by relying on "trade credit." Some suppliers will offer extended payment terms or even ship on consignment to preferred customers. Abuse of this courtesy will mark you as an undesirable credit risk.

Long-term Capital

Long-term capital is used for *capital expenditures*. It is money used to buy fixed assets such as major equipment or buildings. Usually the security for the loan is the item purchased. Long-term loans are characterized by a regular rate of repayment over a longer period of time.

There are many sources of long-term capital. Investment corporations, commercial banks, life insurance companies, mortgage lenders and government agencies are typical.

Collateral

Collateral describes something that ensures repayment—either in full or in part—in case your loan payments are not made. There are many forms of collateral.

A *signature* loan is unsecured. There is no collateral for the loan other than the legally enforceable promise. A signature loan is just a contract and gives the lender no advantage over other contract claimants if there is a default. Few commercial loans are made on a signature only.

A *savings account* is a good form of collateral. You assign your savings account to the lender and turn your passbook over to him. In return, you are given a dollar-for-dollar loan on the amount of the account. Generally, your money stays in the savings account and draws interest. When the loan is repaid, it is signed back to you with the interest intact.

The *cash surrender value* of a "whole life" life insurance policy can be used as collateral. Usually you can borrow directly from the insurance company at favorable interest rates. But the policy can also be used by the bank to secure your loan if you prefer. When the loan is repaid, the policy is signed back to its original condition.

Real estate is a common form of collateral. It's generally used for long-term loans on the property which is the security. Before accepting real estate as collateral, the lender will want to be sure the property is worth more than the amount loaned and has insurance coverage adequate to secure the debt.

Chattel mortgages are like real estate mortgages. But the security is something other than land and buildings. Material and equipment are called chattels. Chattel mortgages are issued for shorter periods of time than real estate mortgages. Like a real estate mortgage, the lender will want to be sure the chattel is worth more than the loan and is adequately insured.

Where To Go For Money

Money for your business will be available from three broad areas:

1. Yourself, your family and friends
2. Banks and lending institutions
3. Government agencies

Depending on your situation, you may use one, two or all three.

Yourself and Others

Very few people have enough money of their own to start a business. It's hard for any wage-earner to accumulate enough excess cash to start an electrical contracting business. But it's been done many times in the past and many people are still willing to make the sacrifices required.

Borrowing from family and friends can create problems. A refusal can mean hurt feelings or a ruined friendship. An acceptance may create problems for you. If your business fails after borrowing $5,000 from Uncle George, you'll still have to see him at Christmas and the annual family reunion. If Uncle George is the unforgiving kind, he may take these semi-annual opportunities to remind you of your shortcomings in the most embarrassing way.

If you have to rely on others, *do it in the most business-like way possible*.

First, look for equity capital—investors rather than lenders. Invite a friend or relative to share the risk in return for a part of future profits. You contribute most or all the effort. Your lender contributes the cash. Offer a fair and sporting proposition. Someone who believes in you and has the spare cash might be willing to bankroll your company and then take a tax loss if it fails.

Most of these arrangements have some kind of control written into them to protect the investor. Even if he doesn't have a voice in the day-to-day affairs, he'll usually want special rights if failure is imminent. There are very few really "silent" partners. But if your lender is an experienced businessman, especially if he's an electrical contractor, his advice may be both welcome and helpful.

If you can't find investors, settle for lenders. My advice is to draw up a loan document just like a bank would. Offer to have a lawyer prepare the agreement if necessary. Spell out exactly what the collateral is and when the loan is to be repaid. Make sure your lender understands the risk involved. You'll be able to live with default much easier if the loan was more like a business deal than a family favor.

Banks and Lending Institutions

The process of filling out the necessary paperwork for a loan was described earlier in this chapter.

Your loan application should include the following, neatly typed:
• Who you are—a short personal biography
• What you want to do—your business objective
• Your estimated starting costs
• Your estimated operating costs
• Your market study
• Your personal balance sheet
• The total loan amount and general repayment terms

Before preparing all this information on your own, *call the bank.* Be sure they make the kind of loans you are requesting and ask for their blank forms. Most will have at least a blank loan application and personal balance sheet.

Information that you submit voluntarily should be just long enough to cover the facts and any positive points. Avoid long, detailed explanations. Be brief but complete. Arrange your pages so each financial table is presented intact on a single sheet if possible. Title each section. Assuming you are typing the application, a typical arrangement might be like this:

Page 1 - Single-space the following information in 3 sections:
   • Name, address and the reason for the loan
   • Personal biography
   • Business objective

Page 2 and 3 - Single- or double-space market feasibility information; double-space number data in columns.

Page 4 - Double-space estimated starting costs

Page 5 - Double-space estimated operating costs

Page 6 and 7 - Single- or double-space personal balance sheet

Page 8 - Double-space total loan amount and repayment terms

If you don't type well and don't have secretarial assistance, hire a professional typing service. It may cost a few dollars, but you're asking for many times that amount. Remember that bankers deal mostly in paper. Neat, attractive paperwork appeals to them.

When your application is ready, call the bank and tell them you want to set up an appointment to discuss your loan application. In some cases, this first contact will simply be dropping off your application and letting the banker have several days to check it out before sitting down to discuss details.

Whatever the arrangements, dress neatly and be on time. You're in competition for the banks lendable funds. They turn down more loan applications than they grant. It's up to you to make them say yes.

The banker will check the accuracy of your market study and estimated costs. He will decide how much he's willing to lend and how much of your capital will be needed. Finally, he'll evaluate your character and business skills. He may also ask that you carry enough business insurance.

If you have a strong personal financial statement and if the lending market is favorable, you don't have to take the first offer. But if your application is rejected, ask for the reason why. Overcome these objections and try again somewhere else. Not all banks follow the same rules and requirements. Don't give up too easily.

Government Agencies

The Small Business Administration both makes and guarantees long term and working capital loans. SBA offices are located in major cities throughout the country and will be listed in the phone book under United States Government.

The SBA has several general rules for qualifying. An electrical contracting business, for example, is considered a wholesale business. It must have expected sales of under $5 million to qualify for help. Also, the borrower must have been turned down by at least two banks or commercial lenders. After that, the SBA will look at your proposal the same way a bank would.

The SBA offers several types of loans. Usually it just provides a loan guarantee, but in the past the SBA has made direct loans. Loans for special situations are also made. To determine your options and what is required, call the SBA and ask for information. They'll send booklets and the necessary forms to get your application started. You may wish to supplement SBA forms with those of your own, as described in the earlier sections of the chapter. This may strengthen your application. But be prepared to fill out plenty of forms and wait for a year or more.

One contractor in the midwest spent a year completing the SBA paperwork and having it processed. When he was finished, the SBA told him "not to get his hopes up, since the money may not be available." Fortunately, after hearing nothing for two months, he got a call that he could pick up the money and go ahead with his plans.

Keep in mind that the SBA looks on itself as a "lender of the last resort." It won't lend money if you have access to other credit. And you must have tried to self-capitalize—to sell a piece of ownership or use your personal credit, if it doesn't impose a hardship.

The SBA puts restrictions on the money once you've received it. But these rules are flexible and can be modified if it will improve operations or prevent a business failure.

Besides the SBA, some state and municipal governments have lending programs that encourage business in a given area. Usually these are only favorable loan terms offered through a private lender. Ask your local banker for information.

Legal Advice

Part of getting into business is putting a lawyer and accountant on your team. You won't need their advice very often. But when you do, both a lawyer and accountant will be worth more than they cost.

If you haven't worked with an attorney before, here's how to select one. Start by asking friends or associates to recommend a lawyer. Your bank may be able to offer several names for consideration. If you don't get some good suggestions, the bar association in your area probably runs a legal referral service. Ask for someone who's familiar with construction law problems.

At this point you don't have any urgent legal problems. That's the best time to select an attorney you can work with. You're not under pressure and can make an informed, rational decision.

Most lawyers are willing to talk to prospective clients for about 20 minutes at little or no cost—especially if you explain that you don't now have legal counsel and expect you'll need some because you're going into business as an electrical contractor.

Tell the attorney that you need some general legal advice on getting started or protecting yourself in business. What you're really doing, of course, is evaluating the attorney. You're looking for someone who's competent in the areas of the law that concern you and can be relied on to serve and protect your interests at reasonable cost.

Interview several lawyers. Compare their answers. Try to picture how you'll be treated when you come back to the office with a real problem. Ask about fees. The most expensive counsel isn't necessarily the best. And beware of any attorney that asks for a retainer at this stage. It isn't necessary or even customary.

Here are some of the topics you might want to cover in your interviews:

Personal Liability - When you are personally liable for company problems—such as on-the-job injuries or financial failure—and how you can be protected.

Company Liability - What your company is liable for and how it can be protected.

Contract Liability - Every contract is unique, but you can get some general advice on where contractors get into trouble, how to protect yourself, and what remedies are available.

Here's a word of advice when dealing with attorneys: Use a lawyer only when absolutely necessary. Learn to handle little disputes and legal problems yourself. Anything involving less than several thousand dollars isn't worth as much as the legal fees may cost. Instead, represent yourself in small claims court or use the American Arbitration Association.

Have an attorney you can call on when needed, but use him sparingly.

How to Protect Yourself and Your Business

This section isn't a substitute for a lawyer's advice. But it provides some guidelines for setting up your business so you're protected against potential problems.

Business Organization

There are three common types of business organization: proprietorship, partnership, and corporation. A proprietorship and a partnership have unlimited liability. This means that the owners are personally responsible for all debts of the company, even in excess of the amount they have invested. In a corporation, only your investment is at risk.

But you may not want to start your business by forming a corporation. It costs several hundred to a thousand dollars plus whatever taxes your state imposes. For one person, it may be best to operate as a partnership or proprietorship until the company is well established.

Figure 10-6 outlines the ways to organize a business and the advantages and disadvantages of each. If you decide to incorporate, you'll probably want your lawyer to handle the details.

Insurance

Every electrical contractor needs several types of insurance. If you have a bank loan, your lender will require adequate insurance.

How much and what kind of insurance to get will depend on the size of your business and what you can afford. As the minimum, you'll want to protect equipment and property from loss by fire, theft and the like. You'll also want to be protected against negligence that results in job-site or work-related mishaps to employees or bystanders. All states require worker's compensation coverage for employees.

Single Proprietorship

Advantages	**Disadvantgages**

Advantages
1. Owner in direct control
2. Low working capital needed
3. All profits to owner
4. Tax advantage to small owner
5. Low starting costs
6. Small amount of outside regulation

Disadvantgages
1. Unlimited debt liability
2. Difficult to raise capital
3. Business ends if owner dies

Partnership

Advantages
1. Easy to set-up
2. Low starting costs
3. Broader management possible
4. Limited outside regulation
5. Two sources of capital
6. Possible tax advantages

Disadvantages
1. Unlimited debt liability
2. Authority is divided
3. Business ends if either partner dies
4. Difficult to raise capital
5. Difficult to find a suitable partner

Corporation

Advantages
1. Transferable ownership
2. Continuous existence
3. Easier to raise capital
4. Limited debt liability
5. Possible tax advantages
6. Specialization of management
7. Possible legal advantages

Disadvantages
1. Most expensive to organize
2. Large amount of record keeping necessary
3. Double taxation
4. Close regulation
5. Subject to possible take-over and/or outside control
6. Charter requirements

Business Organization
Figure 10-6

This coverage is the bare minimum. You may need business interruption insurance, life insurance, contract bonds, automobile insurance, health insurance, completed operations liability insurance and many others. No electrical contractor has the time to become an expert in all forms of insurance he may need. Instead, select an insurance agent or broker like you selected an attorney, someone who's competent and responsive to your needs. Then give that agent or broker enough of your business so that he recognizes you as a good and valued customer.

But don't over-insure. You want enough protection but you don't want to be "insurance poor."

Shop insurance coverages and costs carefully. Premiums do vary.

Government Regulations

All businesses, no matter how small, are regulated by government. Generally, the smaller your business, the fewer the regulations.

You can't begin work without a building permit in most communities. You can't even work as an electrician in many communities without an electrician's license. Figure 10-7 summarizes the most significant government regulations you'll be concerned with and shows where to go for more information.

Other Sources of Advice

Good legal advice is expensive. But there are many ways to get business information that cost little or nothing. Here are a few of the many sources.

Small Business Administration

Besides making loans and loan guarantees, the SBA offers free management assistance to anyone in business or planning on going into business. Ask

Regulation	Employers Affected	Basic Requirements	For More Information
Federal Income Tax	Employers and corportions with active cash flow; profit and loss. Any employer who pays wages to employees (includes family members).	File Declarations of Estimated Income. Withhold taxes from employees' paychecks according to exemptions and applicable withholding rate. Deposit amounts in bank at periodic intervals.	Local Federal Internal Revenue office. Also, write to Internal Revenue Service, Washington DC, 20224 for **Employers Guide & Mr. Businessman's Kit.**
Federal Unemployment Tax	Employers with one or more employees during 20 or more weeks annually.	Contribute required percentage of employees wages and deposit. File Form 940 on or before January 31 annually.	Local Federal Internal Revenue office.
Federal Wage & Hour Laws (Social Security)	Employers who pay over $50 per quarter in wages.	Obtain social security numbers of employees. Withhold required percentage and deposit.	IRS **Employer's Guide** (see above). Local social security office.
Title VII, Civil Rights Act—Equal Employment Opportunity	Employers who pay over $50 per calendar quarter (3 months) in wages.	Treat employees and job applicants equally and fairly with no discrimination as to race, color, religion, sex, age or origin.	U.S. Dept. of Labor, * Office of Federal Contract Compliance Programs. Local Affirmative Action offices.
Occupational Safety & Health Act (OSHA)	All employers except family-owned and operated farms and self-employed persons.	Work site must be kept free of hazards.	U.S. Dept. of Labor, OSHA. Local OSHA.
Employee Retirement Income Security Act	Most employers with employee benefit plans.	Requirements vary depending on plan type.	U.S. Dept of Labor, ERISA Administration. Local office of U.S. Dept. of Health & Social Services
State Income Tax	Varies by state.	Varies by state but are similar to federal requirements. Withhold tax monies and and deposit.	Nearest state Revenue office.
State Unemployment Insurance	Varies by state.	Varies by state but similar to federal requirements. Contribute and withhold suitable wage percent and deposit.	Nearest state job service/unemployment office.
Workers' Compensation Insurance	Varies by state.	Cover employees for possible job-related injuries by state insurance or approved private plan.	Nearest state workers compensation office. Private insurance agent.
State/City/Town Licensing	Varies by area.	Filing of information to insure technical skill.	Local building inspector. State/local consumer protection agency.

*U.S. Dept. of Labor, 200 Constitution Ave., NW., Washington, DC 20210

Basic Government Requirements
Figure 10-7

for an appointment at your nearest SBA office. To avoid confusion, contact them first by phone and explain clearly what your plans are. Make sure you qualify for planning assistance before setting aside time for the session.

Any electrical contractor can benefit from SBA's free management assistance publications. Get a list of publications available by requesting form SBA 115A from:

U.S. Small Business Administration P.O. Box 15434, Fort Worth, TX 76119

Other Low Cost and No Cost Sources
Most large universities have extension departments that offer courses in business planning, computers, technology and other subjects. Look in your phone book under the name of the college or university. In rural communities look in the phone book under the county name for the Agricultural Extension office.

Your public library will have many useful reference materials. Many banks sponsor seminars for the managers of small businesses.

The National Electrical Contractors Association

NECA is the national organization for electrical contractors. Headquartered in Washington, D.C., it also has branch offices in many cities.

Services to members include a large library of published material as well as educational programs. Workshops and seminars are conducted throughout the year in many locations. Topics covered include both basic and advanced bookkeeping, marketing, estimating, customer relations and business operations. For a complete listing of NECA services, ask for Index number 2035 from:

Public Relations Services
National Electrical Contractors Association
7315 Wisconsin Ave.
Washington, D.C. 20014

Membership is available only to operating electrical contractors and costs about $500 per year, plus an assessment based on the number of employees. Most members feel the money is well worth it. For more information about joining, contact the office nearest you or write the national headquarters.

Record Keeping

Every electrical contractor needs to keep good records. If your company is small, you or your wife may be able to handle the chore. Eventually you may have to hire a part-time bookkeeper or train someone to maintain the financial records. Many electrical contractors have a data processing service or small computer that maintains their receivables, payables, payroll and general ledger.

Larger electrical contractors can afford to hire full-time bookkeepers to do their paperwork. They maintain good records because someone has accepted that task as his or her primary responsibility. The smallest electrical contractors have very few records to maintain and can survive with almost any system. It's the contractors who fall between these two groups that tend to have barely adequate or inadequate records.

And understand this clearly: The I.R.S. requires that you keep documents showing your expenses. Failure to do so can result in a major tax liability.

If your operation is big enough to produce many receivables, payables, and payroll documents, to need financial reports, and to file tax returns, you need both a good system of record keeping and occasional help from an accountant.

Here's how to get set up right: Sit down with a qualified accountant. It doesn't have to be a C.P.A., but you may feel more comfortable if he has those initials after his name. It just means he is licensed by the state to do what he does. Ask him to recommend a system and have him explain how it works.

Many small businesses use *single-entry* bookkeeping. Basically, the single-entry system uses the business checking account as the central control. Later, as your sales volume increases, you may wish to go to the double-entry system and shift more of the load to the accountant.

Three rules should guide you when you do your share of the bookkeeping: be accurate; be complete; and be on time. Records that are not processed promptly tend to become confused. Set aside time each day for record keeping. Even if you can afford to hire someone to oversee company records, you'll still have to supervise the work.

No one needs to know more about your business than you do. Recognize that the financial history of your company is being written every day in the documents your business generates. Organize those documents so they can be understood and then take the time to understand them.

One other rule—keep your personal finances separate from the business. Draw a salary like any other employee. Put your paycheck in your personal account and go from there. Never mix your personal finances with the business.

Bookkeeping Terms

The following information may be helpful when talking with the accountant about your bookkeeping operation.

Assets - Anything your business *owns*. Cash, materials, equipment, land, buildings and money owed you by customers (accounts receivable) are some of the assets typically found in a construction business.

Liabilities - Anything your business *owes*. These might be wages owed, taxes owed, money owed to suppliers (accounts payable) and money owed to lenders (notes payable).

Equity - Also called capital, equity is the difference between what the business *owes* and what it *owns*. It's the value left over and belongs to the owner(s).

Bookkeeping Equation - Assets = Liabilities plus Equity. This formula is called the bookkeeping equation. It is simple to keep in mind and is the backbone of all bookkeeping records. Summary statements such as balance sheets and profit-and-loss statements will always show the bookkeeping equation.

Daily Summary - Company financial records start with all the different size sheets of paper used for each transaction—receipts, check stubs, petty cash slips, bank statements and billing invoices. It is important to have a written record—however informal—of all transactions when they take place. If you do a large volume of business, you'll want to enter these transactions on a daily summary sheet. In a smaller business, it may be more convenient to put all the individual records in a folder and summarize them once a week. Be sure to include a record of all transactions. Don't let them go unrecorded longer than one week.

The Journal - Information from the summary sheet is brought together in a journal. Each journal entry shows (1) the date of the transaction, (2) the amount, (3) a brief description, and (4) the name of the account affected by the transaction.

The Accounts - Accounts are a record of the increase or decrease of any type of asset, liability, capital, income or expense. Petty cash, repairs and maintenance, equipment, payroll, taxes and inventory are just a few of the accounts found in a contracting business. Your accountant can help to determine what account categories you need.

Debits and Credits - A *debit* is an account entry representing an increase in value for that particular account. A *credit* represents a decrease. For example, when you buy a load of conduit from a supplier, the money is entered as a credit in your cash account—that is, it's subtracted from the cash balance. The same amount is entered as a debit in your materials account—since that account's value has been increased by the same amount.

By tradition, debits are entered in the left-hand column and credits in the right. When credit entries total more than debits, the account is said to have a credit balance. When the opposite is true, the account has a debit balance. At any time, the total debit balance of all accounts should equal the total credit balance.

The Ledger - This book is where any number of accounts are kept. Since the account entries are debits and credits, the totals of the ledger columns should balance if the bookkeeping is free from errors. This adding and check of the debit and credit ledger columns is called the *trial balance*. It can be done on a monthly, weekly or even a daily basis, to check for errors before they are buried too deeply.

Financial Statements - Your journal and ledger alone would give you the condition of your business at any time. However, the accounts in your books change daily. It is hard to get an overall picture. So, a summary of information is done on a regular basis using financial statements.

The two main financial statements are the balance sheet and the profit-and-loss. The balance sheet is usually done once a month. It summarizes assets, liabilities and equity to show the financial condition of your business. It is called the balance sheet because it follows the bookkeeping equation—assets will equal liabilities plus equity.

The profit-and-loss statement shows the income and expenses of your business over a given time. P&L's are usually done on a quarterly or yearly basis. They show at a glance your profit or loss for that time period.

Most small businesses use a combination of in-house bookkeeping and an accountant. For example, you may want to keep your own daily summary, journal posting and accounts, and your own payroll. You can then hire the accountant to do a trial balance, a P&L and prepare tax reports at regular intervals. As your business grows and changes, your accountant will recommend where new account categories are needed.

Bookkeeping supplies are sold at most better office supply stores. Many different forms are available. Select those recommended by your accountant or what you find works best. Besides your record set, you may want to buy a file cabinet or lock-box to keep them in. A desk-top calculator is nearly essential.

If you don't have experience with small computers, don't buy a computer to keep your books. Microcomputers are great, but learning to use one can take weeks. Unless you have a computer, and understand how to use it, and are looking for some project it can tackle, stick to electrical contracting.

Inventory

Here are some ways to cut your inventory burden:
• Favorable credit terms offered by eager suppliers have helped many new electrical contractors.
• Always ask about quantity discounts. Everything is available at a discount if you buy enough.
• Have materials delivered directly to the job and at about the time when it will be needed. It cuts your risk of loss and reduces your payables for work in progress.
• Plan to protect everything delivered to the job from both weather and theft as soon as it arrives. Some insurance policies will require that it is kept in a locked enclosure before the insurance will apply. This can be your own building or a temporary enclosure on the job-site.
• Set up an inventory control system. It's not complicated. A sign-out sheet on a clipboard next to the door of the secured area will work for most operations. A trustworthy person should have a key and be responsible for checking materials in and out. The sheet should have columns for the

date, material description and serial number, the amount, job location and the person removing the materials. Compare the sheet with the master inventory and note any shortages. A double-check is to compare it with a specific job schedule. For example, if you see a lot of finish materials signed out for a job that's in the roughing stage, it's obvious you have a problem.

Advertising

Most electrical contractors are content to limit their advertising to the telephone yellow pages. Most work comes from trade contacts. But many alert electrical contractors are more aggressive. They find it pays to advertise. Try it yourself and see.

If you're thinking of advertising, plan to target your ads where they'll do the most good. Radio, print and TV advertising all appeal to different markets. Their ad salespeople will be able to tell you exactly who you can reach through their media.

For example, a radio commercial on a Top 40 AM rock-and-roll station is a waste of your money. But a small ad in a quality homeowners magazine may bring some remodeling work.

There's no sure way to know what the results will be before you place the ad. But here's a tip: If you see some ad for an electrical contractor again and again in a local paper, assume that it's paying off for him and could for you too. Try it. Keep track of the results. If it doesn't work, try something else until you find something that does.

Advertising can be expensive. And it may be hard to see any direct benefit. And no advertising does anything more than find prospects. Don't expect to run an ad and then sit back and write up the work orders. You still have to make the sale.

Here's one good and low cost way to put your name before the buying public: Place your company ad in the program or pamphlets put out by local organizations or civic groups sponsoring an event. It will reach only a limited audience, but the cost is very low. They reach all local customers, and there's the added plus of being associated with an event that benefits the community.

Be aware that many ads in magazines, newspapers and on the radio are bartered, not paid for with cash. If an advertising company needs your services, offer to do the job and be paid in advertising equal to the dollar value of your work.

Professional Organizations

Anyone who runs a business should consider the advantages of joining local business clubs and civic groups. Nearly every community has business associations. National organizations such as Kiwanis or Lions probably have a local chapter in your area.

Membership in these organizations is a good way to exchange ideas with others in business. Membership makes you aware of possible business opportunities. People like to do business with people they know and respect. There's an advantage to being known in the community where you operate.

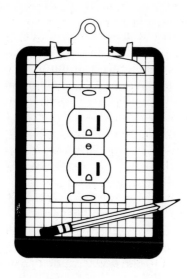

Chapter 11:

Estimating and Bidding

Estimating and bidding are key parts of every electrical contracting business, whether large or small. A few contractors are lucky enough to get work from regular customers on a non-competitive basis. But most electrical contractors get work only when they submit the low bid.

The difficulty with bidding should be obvious: Bid too high and you get nothing; bid too low and you're always busy but never making any money. It doesn't take very many of either kind of estimates to put most electrical contractors out of business.

The following sections explain how to keep your estimates from being either too high or too low.

Costs, Overhead and Profit

All electrical estimates have three parts: cost, overhead and profit. While they may not be identified under those headings, they must be in every estimate somewhere. Otherwise you're not going to stay in business for very long.

Cost is divided into two parts—*material* cost and *labor* cost. These are sometimes called *direct job costs*. They are the cost of materials needed for a specific job and the cost of labor to install those materials.

You'll hear some contractors say that their material-to-labor cost ratio is 60/40. That is, every $60.00 they spend on materials costs another $40.00 in labor for installation. This figure has been used for many years and probably applies within 5 or 10% on many jobs. But it doesn't consider your hourly labor cost, the productivity of

your crews or the types of work you do. Your estimates have to be based on more precise costs.

On larger jobs, your estimates will include a figure called *indirect job cost* or job overhead. This is job related expense other than labor and materials. Jobsite insurance, special equipment rental, job office, job telephone and the like fall in this category. Job overhead should not be confused with normal business overhead. Job overhead will be about 8% of the direct cost on many jobs. Calculate how much your job overhead is on the work you do and add it to the total job cost before completing your estimate.

Business or operating overhead, usually spoken of simply as *overhead*, is the cost of running your business, even if you have little or no work in progress. Overhead costs commonly include rent, telephone, advertising, utilities, car and truck expense, office salaries, office supplies, your salary as manager of your business, interest on loans and legal fees.

For many electrical contractors, their business overhead will be about 25% of total job cost. It's added to the total job cost just like job overhead.

Profit is the owner's earnings on the money he has invested in his business. Usually about 10% is added to the total cost in each estimate as profit. But profit is dictated by job and competitive conditions. There is no single figure that will fit all jobs.

Estimating

Good estimating is an essential part of any successful contracting business. Every estimate is a guess, but a very well-informed guess. When done carefully, it will accurately reflect actual project costs.

Your estimates are a written record of expected job costs. They aren't legally binding on your company, but they are the basis for your bid which may become part of the written or oral contract.

Estimate Forms
You'll reduce errors and prepare better organized, more complete estimates if you use good estimating forms and follow the same procedures every time. This chapter includes many estimating forms for your use. If you have better forms and find they work well for you, use your own forms. If the forms in this book are better, order a supply from your local office supply store, or from:

The National Electrical Contractors Association
7315 Wisconsin Avenue
Washington, D.C. 20014

No two estimators or contractors use the same estimating system. In fact, there is no one right way to figure job costs. But there are many points that most estimators can agree on: The forms you use should be tailored to the work you estimate. Neatness counts. A neat, well organized, logical estimate makes it easy for someone else to check your figures. And when the estimate has to be revised, as most estimates do, revision will be easier.

Record all your calculations. Have a form with enough columns to show how each figure was developed—and use each column when needed.

Use checklists and standard procedures to avoid missing any cost item. Your estimate on anything you miss is always zero. That's a 100% error!

Listing each item of material needed on your *take-off sheet* is the first step in every estimate. You read the plans and specifications and systematically list each item needed.

The take-off sheet will have columns for written or symbol descriptions of the materials. In the next column you list quantities. A good take-off sheet has plenty of columns. This makes it easy for you to transfer symbols and picture information from the plans and specifications. Figure 11-1 shows five typical take-off sheets.

No dollar values are put on the take-off sheet.

When the take-off is done, materials that have the same cost are grouped and listed on the *pricing sheet*. See Figure 11-2. A cost for each is found and placed in the material cost column. This cost is multiplied by the quantity needed. This total goes

in the material extension column. For each item of material, you need an estimate of the labor required for installation.

Labor costs can be given either in time or money. For example, assume that an electrician gets $27.00 per hour. This is 45 cents a minute. Suppose you know that installation of a single-pole circuit breaker will usually take him three minutes. The labor cost per item will be 3 x 45 cents, or $1.35. This figure goes in the labor cost column. If eight of these breakers are needed, the $1.35 is multiplied by eight and $10.80 is placed in the labor extension column.

Some contractors, however, prefer to figure labor in time rather than dollar cost. In this case, the labor column would be labeled manhours. The cents column would be used for minutes and the dollar column for hours.

As a general rule, all similar items that cost the same and are installed in the same amount of time can be combined under one heading on the pricing sheet. But be alert to separate items that are alike but have different material or installation costs. For example, the take-off sheet may show twelve different listings for No. 12 wire. The total footage and price can be combined on the pricing sheet. However, the same sheet may show listings for both EMT and rigid conduit. These should be listed separately. They'll have different costs.

Large job estimates will have an *estimate summary sheet* like that in Figure 11-3. This sheet summarizes information on the pricing sheets. It will include listings for job overhead, labor cost breakdowns, material handling and other like items. The estimate summary shows at a glance the total dollar values for materials, labor, overhead and profit. It will also show any special material and labor costs.

The *cost analysis sheet*, like the estimate summary, is not used on every job. The analysis sheet provides another way of looking at the information on the pricing sheet. Some parts of a job will cost more because special materials are required or installation is more difficult. The cost analysis sheet is a tool for examining this part of the job. These sheets are also useful for comparing estimated costs with actual costs as the job progresses.

All electrical contractors, large and small, should keep *job comparison records*. This becomes your best source of information on labor productivity for your crews. When your estimate is complete, list the estimated labor cost for each part of the job. When actual labor costs are known, write the hours in on the appropriate line. See Figure 11-4. Add notes that explain why the estimate dif-

SWITCHBOARD-PANEL BOARD
AND FEEDER SCHEDULE

FORM 10-E
Reprint from the
NATIONAL ELECTRICAL CONTRACTORS ASSOCIATION, INC.

ESTIMATE NO._____
SHEET NO._____
OF_____SHEETS
DATE_____

JOB _____

ESTIMATED BY _____

LOCATION		SWITCHBOARD						PANEL BOARDS					RACEWAY AND FITTINGS											WIRE AND CABLE				
		TYPE_____ VOLTS_____ MAINS_____						TYPE_____ MAINS_____ MTG._____																				
				FEEDER CIRCUITS		MAINS			PANEL	BRANCH CIRCUITS																		
FEEDER NO.	FROM	TO	NO.	TYPE	POLES	AMPS.	FUSES	NO.	15A	20A			TYPE	SIZE	LENGTH	ELBOWS	BENDS	PULL BOXES	FITTINGS	HANGERS	TER-MINALS	NIPPLES	TRENCH	SIZE	TYPE	NO.	LENGTH	TOTAL

Typical Take-Off Sheet
Figure 11-1

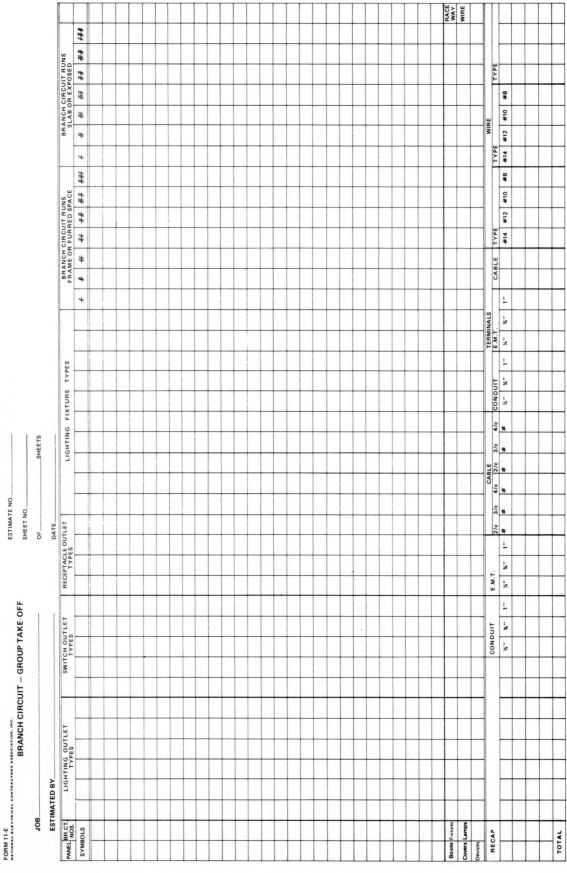

Typical Take-Off Sheet
Figure 11-1 (continued)

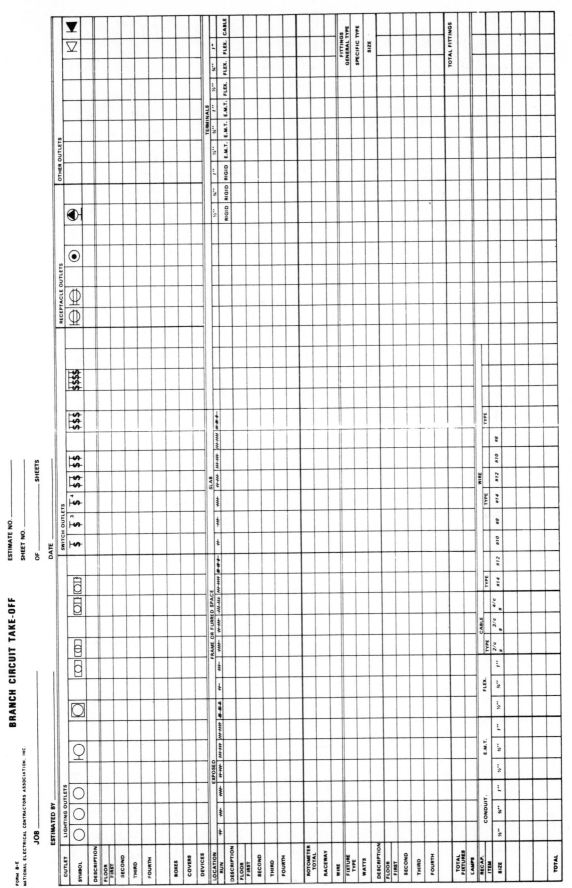

Typical Take-Off Sheet
Figure 11-1 (continued)

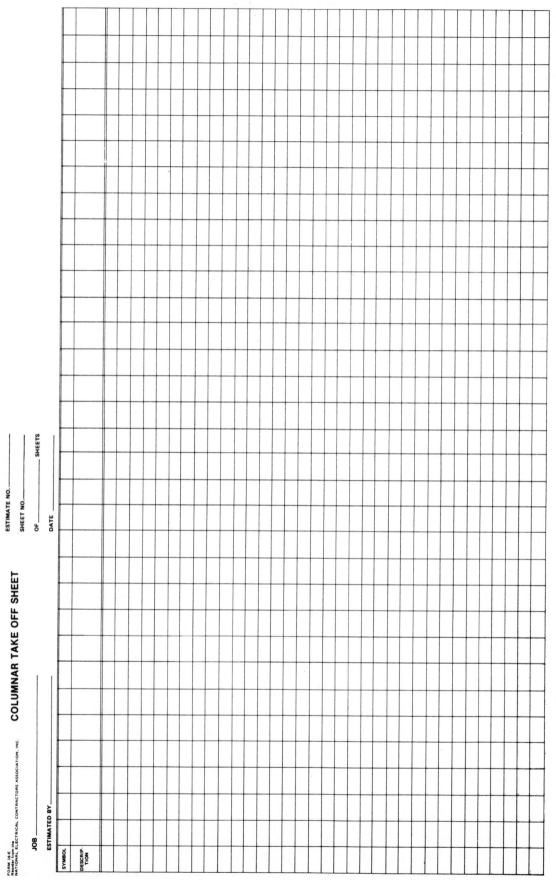

Typical Take-Off Sheet
Figure 11-1 (continued)

FORM 9-E
NATIONAL ELECTRICAL CONTRACTORS ASSOCIATION, INC.

POWER SCHEDULE

ESTIMATE NO. _____

SHEET NO. _____

OF _____ SHEETS

DATE _____

JOB _____

ESTIMATED BY _____

Ct. No.	Location or Use	MOTOR OR OTHER POWER OUTLET					EQUIPMENT				RACEWAY AND FITTINGS									WIRE				TOTAL
		Size	Volt	Phase	Type	Dis-connect	Br. Ct. Protec.	Starter	Control	Size	Type	Length	Fit-tings	El-bows	Term	Nip-ples	Flex	Flex Conn.	Size	Type	No.	Length	

Typical Take-Off Sheet
Figure 11-1 (continued)

FORM 22E NATIONAL ELECTRICAL CONTRACTORS
ASSOCIATION, INC.

ADP CODING AND PRICING SHEET

JOB _____ ESTIMATE NO. _____

WORK _____ SHEET NO. _____ OF _____ SHEETS

ESTIMATED BY _____ PRICED BY _____ EXTENDED BY _____ CHECKED BY _____ DATE _____

JOB SEC. CODE			ITEM CODE						*LC	MC	QUANTITY							MATERIAL DESCRIPTION	MATERIAL PRICE	U N I T	MATERIAL EXTENSION	LABOR UNIT	U N I T	LABOR EXTENSION
1	2	3	4	5	6	7	8	9	10	11	12	13	14	15	16	17	18							

*LC-LABOR CODE, MC-MATERIAL CODE

Pricing Sheet
Figure 11-2

PRICING SHEET

ESTIMATE NO.

JOB _____

WORK _____

Sheet No. _____ Of Sheets _____

ESTIMATED BY	PRICED BY	EXTENDED BY	CHECKED BY	DATE

	✓	MATERIAL	QUANTITY	MATERIAL PRICE	PER	MATERIAL EXTENSION	LABOR UNIT	PER	LABOR EXTENSION
1									
2									
3									
4									
5									
6									
7									
8									
9									
10									
11									
12									
13									
14									
15									
16									
17									
18									
19									
20									
21									
22									
23									
24									
25									
26									
27									
28									
29									
30									
31									
32									
33									
34									

Totals This Sheet Transferred To Recap By-INITIAL
Form 4E
Reorder from National Electrical Contractors Assoc.

MATERIAL ▶   LABOR ▶

Pricing Sheet
Figure 11-2 (continued)

Pricing Sheet Nos.	Section of the Job	Code	Standard Materials Cost	Special Materials ①	Labor Man Hours Non-Factored	Job Factor + or −	Job Factor MH or % ②	Non-Prod. Labor Factor + or −	Non-Prod. Labor Factor MH or % ③	Productivity Factor + or −	Productivity Factor MH or % ④	+ or − Net Labor Hour Adjustments	Labor Man Hours Factored	Estimating Comments Relative to Construction Phases
_to_														
_to_														
_to_														
_to_														
_to_														
_to_														
_to_														
_to_														
_to_														
_to_														
_to_														
_to_														
_to_														
_to_														
_to_														
_to_														
_to_														
	Miscellaneous Material													
		Totals	$	$										

DIRECT JOB EXPENSES:	DOLLARS
Insurance Add'l For This Job	
Insurance—Other	
Insurance—Other	
Permit and Inspection Fees	
General Supervision ⑥	
Job Site Engineering and Drafting	
Storage & Job Shed	
Job Time Keeper and/or Stockman	
On Job Truck Expense	
Job Telephone	
Temporary Facilities ⑦	
On Job Tool & Equipment Costs	
Travel Expense	
Subsistence	
Sub Contract	
Labor Adder Dollars ——→	
Other DJE	
Other DJE	
Other DJE	
Other DJE	
Total Direct Job Expenses $	

LABOR ADDER % ANALYSIS ⑤	%
N.E.I. Fund	
NECA Service Charge	
Social Security	
N E B F	
Unemployment Insurance	
Workmen's Comp. Ins.	
P.L. & P.D.	
Local Health & Welfare	
Other	
Other	
Total Labor Adder %	
Labor Adder Calculation:	
Total Labor Dollar Cost $	
Labor Adder %	x
Labor Adder Dollars $	
(Insert as a Direct Job Expense)	←——

Estimator's Evaluation of Job, Non-Prod. & Prod. Factors	
Job Factor Defined	Non-Prod. Labor Factor Defined
1. Type of Building	1. Craft Coordination
2. Working Conditions	2. Study Time
3. General Contractor	3. Material Handling
4. Electrical Contractor	4. Lost Time

Notes by Estimator:

ESTIMATE SUMMARY	DOLLARS
Standard Materials at Cost	
Special Materials at Cost	
Labor Man Hours Factored ________	
Average Rate Per man/hr. $________	
Total Dollar Cost—Direct Labor (Lab. MH Factored x Ave. Rate)	
Total Direct Job Expenses	
PRIME COST	
Overhead (Based on Prime Cost) ⑧ ______ % X $________ Prime Cost OR	
O.H. (Std. Mat., Labor and DJE) ⑧ ______ % X $________ PLUS O.H. (Spec. Mat) ______ % X $________	
TOTAL JOB COST	
Net Profit Based on ______ % of Total Job Cost	
SUB-TOTAL OR SELLING PRICE $	
*Sales and/or Excise Taxes	
**Payment & Performance Bond	
**Interest on Financing	
TOTAL SELLING PRICE $	

*Can be Included in Material Cost
**Can be Included in Direct Job Expenses

CALCULATE TO THE NEAREST DOLLAR AND WHOLE MAN HOUR.

Courtesy: National Electrical Contractors' Association

Estimate Summary Sheets
Figure 11-3

Fig. 11-4   A Job Comparison Sheet

Labor Cost Summary - Hillcrest Apts. Job

Job Code	Type Of Work	Estimated Manhours	Actual Manhours
01	Roughing in Branch Circuits	1,590	1,720
02	Feeder Circuits	349	305
03	Branch Circuit Wiring	471	431
04	Switches and Receptacles	115	125
05	Feeder and Service Cable	124	120
06	Lighting Panels	105	115
07	Power Panels	75	70
08	Meter Wiring	38	36
09	Motors	145	160
		2,912	3,082
	Foreman - Berger	810	695
	Total Manhours	3,722	3,777

#01 - New man on Roughing crew.

#02 + 05 - Good job by Jenkins + Smith!

Berger - Spread too thin. Was starting Sunnyview job when Motors + Light Panels were installed. Jenkins as foreman if jobs increase?

Job Comparison Sheet
Figure 11-4

fered from the actual cost, if it did, and suggest how the estimate could have been improved. These sheets are very valuable when estimating future jobs.

Plans and Specifications
Chapters 5 and 6 explained how to draw up plans to do the actual wiring. Later in this chapter we'll explain in detail how to make a material take-off from the plans and specifications. This section tells you how the plans and specs fit in the estimating and bidding process.

Your take-off sheet, when completed, is an organized list of all materials to be installed on the job. The source of this information is the plans and specifications. You go over every symbol and line in the plans and every word in the specs to find anything that will add to your cost of doing the job. Completeness is essential. Underestimating a few feet of wire or conduit won't cost you more than a few dollars. But missing a room or entire floor or one sheet of the plans can be a very expensive miss.

The job specifications are as important as the plans for your material take-off. You don't really know what the job requires until you've read the specs. They identify the quality of materials and workmanship required and explain the construction procedures to be followed.

Since 1966, United States construction industry has followed the Construction Specifications Institutes' *Format For Construction Specifications*. The CSI Format has sixteen *Divisions*:

Division
  1 - General Requirements
  2 - Site Work
  3 - Concrete
  4 - Masonry
  5 - Metals
  6 - Wood and Plastics
  7 - Thermal and Moisture Protection
  8 - Doors and Windows
  9 - Finishes
10 - Specialties
11 - Equipment
12 - Furnishings
13 - Special Construction
14 - Conveying Systems
15 - Mechanical
16 - Electrical

Each division is further divided into sections for each type of material or equipment used in construction. You will be most concerned with Division 16.

Figure 11-5 is the index to Division 16 taken from the specifications for a commercial project. It's common to have an index like this if the specifications for any division are more than a few pages long. The index makes finding the right specification much easier.

Under C in Figure 11-5 you'll see a list of 5-digit numbers followed by a description. These are the sections under Division 16. The first two digits are all 16, indicating that all sections fall in Division 16.

Figures 11-6A and 11-6B are typical specifications for sections 16105 and 16106.

Follow the specifications if you expect to get paid and want to stay out of court. There's no ex-cuse for installing the wrong material or installing it the wrong way.

You can substitute materials if permitted by the specifications. If substitution is allowed, the architect or owner usually reserves the right to approve the substitute. On a larger job the architect will require that shop drawings, test reports, manufacturers' specifications and maintenance and operating instructions be submitted for approval. This prevents the installation of cheap or shoddy materials or equipment that may not fit the overall building design.

Fixture, material and equipment substitution can be important when estimating. If you can get approval for a fixture that costs less than the one specified, you have a competitive edge. But don't make a pest of yourself by asking for approval of cheap material without regard to quality. That would damage your reputation as a quality contractor. An electrical contractor known for poor quality work will probably be rejected outright by the owner or general contractor, in spite of having the low bid.

Get approval for material and equipment substitution as soon as possible after receiving the specifications. The take-off isn't complete until you know the materials to be estimated.

Changes to the original plans are called *addenda*. Construction is too permanent and costs too much to overlook a mistake discovered in the plans before construction begins. Also, the owner may have some last minute changes Addenda are usually sent out after the plans are distributed but before bids are submitted. Your bid should indicate that the quote covers work described by the plan and specifications and addenda. Identify which addenda are covered. For example, if three addenda were received, your statement might be: "Price includes all addenda up to and including Addenda No. 3."

Submitting the low bid isn't enough if you don't state which addenda are included in the figures.

Any changes in the plans or specs after the contract is awarded will be through *revisions and change orders*. These will reflect changes needed to comply with building codes, to simplify the work or to correct architectural details. All change orders and revisions should show the following:
  • A description of the change
  • Who will do the work
  • Who will pay for the change
  • Who is authorizing the change
Change orders should be in writing and approved before you do the work. They protect you by ensuring that you're paid for work not included in the estimate. But be careful about estimating the cost

16010 General Provisions

16011 Work Included

A. Furnish all materials, equipment, labor, and related items required to complete work indicated on drawings and/or specified. Attention is directed to the **"General Conditions"**, **"Supplementary Conditions"** and other Divisions of the Specifications which affect the work under this Division.

B. The work under this Division includes but is not limited to the following items.

1. Power and lighting wiring system in building and on the site.
2. Temporary electrical work.
3. Exit lighting and emergency lighting system
4. Lighting fixtures.
5. Telephone conduit system.
6. Fire alarm system.
7. Intercom system.
8. MATV system.
9. Emergency call system.

C. **Index**

16010 General Provisions
16012 Related Work Specified Elsewhere
16014 Permits and Inspections
16100 Basic Materials and Methods
16101 Equipment Identification
16104 Conduit
16105 Outlets and Junction Boxes
16106 Power Outlets
16109 Wire and Cable
16110 Wiring Devices
16111 Safety Switches
16113 Overcurrent Protection and Grounding
16114 Wall Plates
16400 Electric Service
16406 Panelboards and Cabinets
16408 Motor Starters
16500 Lighting
16600 Special Systems
16702 Telephone Systems
16703 Fire Alarm Systems

Specification Index for Division 16
Figure 11-5

of changes. Your overhead, material, labor and equipment costs will almost always be higher per unit installed under a change order than if the same item was included in the original plans. And note that changes reducing the amount of work done may reduce your labor, material ana overhead costs very little or not at all.

Subcontracting
On most jobs you'll be a subcontractor rather than the general contractor. Your bid is to a general contractor who contracts with the owner. The general contractor checks your bid for accuracy and then uses it to make up his own bid. If he gets the job, he'll sign separate contracts with the subcontractors.

160106   **Power Outlets**

A.   Furnish and install outlets for, and make connections to all power operated equipment indicated on the drawings. (See Plumbing and Heating drawings).

Make final electrical connections to equipment and wire control circuits. Set dryers and ranges in place, make final connections, and put into operation.

B.   All measurements shall be taken from the Architect's general construction drawings except when dimensions are shown on electrical plans, then such dimensions will be used. All junction and pull boxes shall be located in utility area walls only where required.

C.   Unless otherwise noted, boxes for the following outlets shall be located with their center lines at elevation above the finished floor line as follows: (See Architectural Plans).

Description	Mounting Heights	
Switches and Emergency Call Switches	3'-0"	Or as noted on Drawings.
Receptacles and TV Outlers	2'-0"	Unless over work counters or baseboard radiation, then as noted or required.
Fire Alarm Stations	4'-0"	Or as noted on the Drawings.
Fire Alarm Horns	6'-6"	Or as noted on the Drawings.
Telephone	---	As noted on the Drawings.
Intercom Station	4'-6"	Or as noted on the Drawings.
All other Outlets	---	As noted on the Drawings.

D.   Switch locations shall be checked at the site to determine that no switches shall be behind doors when doors are in open position.

E.   The Owner reserves the right to change the location of any outlet prior to installation. Said changes to be located in such a manner as to cause no additional cost to the Electrical Contractor.

F.   The Electrical Contractor shall cooperate with others to locate ceiling outlets so that lighting fixtures shall conform to the pattern of the ceiling. Where conflict occurs between recessed fixtures, ceiling and beams or other obstructions, the contractors involved shall consult the Architect for proper location.

Typical Specification Pages From the Electrical Section
Figure 11-6A

These separate contracts establish the legal relationship between the contractor and his subs, but don't bind the owner or architect. But the owner or architect usually reserve the right in their contract with the general contractor to reject a proposed subcontractor. The owner probably has protected himself with a provision in the general contract which limits his liability if a dispute arises between the sub and the general contractor.

Contracts

The contract binds the parties in a mutual agreement. Your obligation is to provide materials and labor in return for the agreed price. On a larger job there may be several separate contracts. On small residential jobs, the contract may be only a signed estimate or an oral agreement.

An informal oral contract is just as binding as a written document. The problem with an oral con-

16105   Outlet And Junction Boxes

A.   Standard 4" octagonal boxes Federal Specifications W-J-800 (complete with fixture stud for lighting with depth as required by Code) shall be used for ceiling outlets except as noted below.

B.   In interior stud partitions, use 1½" deep boxes and ½" raised covers. Outlets occurring back-to-back within the dwelling unit shall have 1½" deep extension ring with ¼" raised covers with receptacle openings offset where necessary for devices specified. Back-to-back outlets shall not be permitted in dividing walls between dwelling units or in fire rated walls and partitions. Stagger all such outlets to reduce noise transmission.

C.   Standard square galvanized steel or plastic boxes complete with proper plaster ring shall be used for all switch, receptacle and special outlets in all walls. Gang boxes shall be used where more than two devices occur at one location or as required by Code. Surface wall boxes shall be 4" square with ½" raised covers.

D.   Sectional type wall boxes or handy boxes shall be used for all switch, receptacle and special outlets in glazed tile, face brick, and unfinished block walls.

E.   Provide special fixture and/or auxiliary support for the outlet box where weight of fixture requires greater support than given by the fixture stud.

F.   Brackets and switch outlets shall clear doors by 6 inches where possible.

G.   All outlet boxes shall be securely fastened in place, independent of support from connecting conduit. All boxes shall be of sufficient size for the number of wires in box and shall fit flush with the final finished wall or ceiling surface in every case. Install metal box extenders and bond to the box where necessary.

Typical Specification Pages from the Electrical Section
Figure 11-6B

---

tract is that the parties may disagree on what the contract terms are.

You may occasionally see a *letter of interest* used to speed up the start of a job. This letter is an authorization by the owner for a limited amount of work. It is as legally binding as a contract for the work it covers.

No matter what type contract, be cautious about agreeing to anything before you understand your obligations. Avoid contracting with anyone that has a poor reputation for performing his obligations. If the other party won't do what he agreed to, your only alternative is legal action. Be sure the contract spells out the exact work requirements, payment schedule and all other items clearly worked out and mutually understood.

No matter how friendly your relationship with the general contractor, disagreements will arise. The contract should spell out how unexpected events will be handled. A good contract can keep misunderstandings from becoming bitter disputes. No one really wins in a lawsuit except the attorneys. A major work dispute will drain the profits out of any job. It creates an atmosphere of bitterness and mistrust that will sour work relationships, perhaps permanently.

Before entering any agreement, remember the following:

• Be sure you understand the job requirements and scope of the work. If you have doubts about certain clauses or provisions, ask questions until satisfied.

• Don't take anything for granted.

• Many contracts are drawn up on standard legal forms. Be sure deletions and additions to these forms are understood.

• Get legal advice if necessary on points you don't understand.

• Final payment should be based on completion of the electrical work and not on project completion. Be sure the payment schedule is included in the contract.

Payment Schedules

The *payment schedule* establishes when payments are due on work completed. For small jobs, you'll probably get only one payment on completion. A larger job will have progress payments that cover a portion of labor and material costs as work goes forward. Five or 10% of the total is held back until the job is inspected by the owner or architect and declared 100% complete.

Payment schedules may be part of the contract documents. More often, they're negotiated separately between the electrical contractor and the general contractor or owner. Try to arrange payments that work to your advantage. This means working with as much of the owner's money as possible.

You may have to sign a *lien waiver* to get paid. These are legally binding documents that verify that you have paid your bills for the job involved. This protects the owner and general contractor from lawsuits for your unpaid bills. A false lien waiver is evidence of fraud and a serious offense.

Your payment schedule should be based on your performance, not tied to someone else's schedule or based on a percent of project completion. As a subcontractor, the electrical contractor will have no control over other contractors.

If payment to the general contractor by the owner is held up because of something the general did or failed to do, you're entitled to payment on demand from the general.

Bids

Types of Bids

A bid is your offer to do the job at the price stated. It's sometimes called a proposal or a quotation. Whatever it is called, it is legally binding for the work it covers when accepted and signed.

Your bid may be subject to rules stated in the specifications or contract documents. If your bid violates these rules, it can be rejected. Often the owner reserves the right to reject bids outright.

Some plans show sections of the work designed two different ways. The owner or architect will decide which one of the alternates will be used. This puts you to the extra work of preparing alternate estimates. You prepare a *base bid* for the complete job, including one way of doing the section. Then you draw up another bid covering the alternate section design. This is the *alternate bid* and is included in the formal bid along with the base bid. Be sure to identify alternate and base bids clearly.

Don't overlook items other than costs that are requested in the bidding instructions. For example, some bids must include items such as work schedules, start and finish times, proof of financial stability, or certificates of insurance.

Bid Forms

Some architects supply bid forms which the contractors must use. Otherwise, bids can be made on a form such as Figure 11-7. Some office supply stores and the National Electrical Contractors Association sell forms like Figure 11-7. Some contractors use the pricing sheet as a bid form.

The bid form usually includes the names of the parties, the date, a work description, a location of the work site, the price for the work, payment rate and a place for the signatures.

Bid Record

A *bid record* is a written record of bids prepared and submitted. If you prepare more than a few bids a month, start a bid record such as in Figure 11-8.

Bid Security

Some owners demand some kind of *bid security* to ensure that you will enter into a contract to do the work at the price quoted. Sometimes a certified financial statement is all that's needed. On other jobs you may have to submit a certified check or post a bid bond with the bid. The amount of the check or the bond is usually a percentage of the total job cost.

Some projects require a performance bond that guarantees you will complete the job. If you don't, the bond may be forfeited or the bonding company may have to hire another contractor to do the work. The bond will cost you from one to two percent of the amount of your bid.

If you have to post a certified check to guarantee completion, be sure the contract provides for return of the check when you have finished work, not when all work by all contractors has been completed.

Bid Invitations and Information

Bids for work are solicited by word-of-mouth, and through ads in construction magazines and local newspapers. These bid invitations usually describe the job and tell where interested parties can go for more information.

On request, you will usually get a bidder's package that includes *bid instructions*. These are the rules under which your bid will be submitted.

FORM 40
NATIONAL ELECTRICAL CONTRACTORS
ASSOCIATION, INC.

UNIFORM PROPOSAL

From...
                    Electrical Contractor

To...   Proposal No...
....   Date...

We hereby propose to furnish all labor and material necessary to install Electric Wiring in the...
...........................located at..
in accordance with the following specifications, and subject to the conditions of contract stated on the reverse side of this sheet.

SCHEDULE OF OUTLETS AND BELL WORK

LOCATION	Ceiling Light Outlets	Bracket Light Outlets	Convenience Outlets Single	Convenience Outlets Duplex	Floor Outlets	Heater Outlets	Range Outlet	S. P. Switches	3-Way Switches	4-Way Switches	Entrance	Special Outlets and Bell Work	Notations

Unit Price for Additions

Branch Circuit Wiring is to be { Rigid Steel Conduit / Flexible Steel Conduit / Armored Cable / Knob and Tube } work, except exposed work in the basement, which is to be.........................
.... Service wires are to be installed in rigid steel conduit. The service
switch and meter loop are to be located in... Wall switches are to be the...........................
....type, and receptacles the...type.

The customer has the privilege of cancelling this contract at any time before the work is started, upon payment of 10% of the price stated below. This proposal is void if not accepted in writing within...........................days after this date.

The price for the work described above will be... Dollars

($.......................) payable on the following terms:

....
....
....

Accepted by...   Electrical Contractor
                    Customer
Date..   By...

Bid Form
Figure 11-7

CONDITIONS OF CONTRACT

1. Wiring Standard—All workmanship and materials are to comply with the requirements of the National Electrical Code and the City Ordinances governing this class of work.

2. Fixtures—Lighting fixtures and lamps are not included in this proposal.

3. Materials—The contractor has the privilege of using any make of material or equipment as indicated in the specifications or shown on plans unless the selections of the material shall have been made (in the specifications) prior to the signing of the contract.

4. Unit Prices—Unit prices submitted with this proposal shall apply only for additional work.

5. Additional Work—Additional work may be ordered by the customer at any time before the work is started, at the unit prices named in the schedule. Any other change from the work as herein described involving additional cost of labor or material is to be paid for by the customer in addition to the contract price named herein, at the price agreed upon or at our regular rates for time and material work.

6. Written Orders—The contractor shall receive written orders for all additional work whether it be on a contract price, unit price or time and material basis, before proceeding with such extra work.

7. Payments for Additional Work—Payments for additional contracts to the original contract shall be made under the same terms and conditions as are embodied in the original contract.

8. Contract Payments—The contractor shall not be required to proceed with the installation of the work if the payments applying on same have not been made as specified in the contract.

9. Unavoidable Interruptions—It is hereby mutually agreed that the contractors shall not be held responsible or liable for any loss, damage or delay caused by fire, strikes, civil or military authority, or by any other cause beyond his control.

10. Transfer of Title—If the customer disposes of the real estate by sale or otherwise, the full amount remaining unpaid on this contract becomes due at once and payable within 48 hours after date of such disposal.

11. Arbitration—Any controversy or claim arising out of or relating to this contract or the breach thereof, shall be settled by arbitration, in accordance with the rules, then obtaining, of the American Arbitration Association, and judgment upon the award rendered may be entered in the highest court of the forum, state or federal, having jurisdiction.

Bid Form
Figure 11-7 (continued)

Form 41

NATIONAL ELECTRICAL CONTRACTORS
ASSOCIATION, INC.

PROPOSAL FOR ELECTRICAL WORK

From _
Electrical Contractor

To _   Proposal No. _ _ _ _ _ _ _ _ _ _ _ _

_   Date _ _ _ _ _ _ _ _ _ _ _ _ _ _ _ _

    We hereby propose to furnish all labor and material necessary to provide the Electrical Installation in the _ _ _ _ _ _ _

_ _located at_ _

in accordance with the following specification, and subject to the conditions of contract stated on the reverse side of this sheet.

The price for the work described above will be

payable on the following terms:

    This proposal is void if not accepted in writing within_ _ _ _ _ _ _days after this date.

Accepted by _   By_ _
             Customer   Contractor

Date _   _

Bid Form
Figure 11-7 (continued)

CONDITIONS OF CONTRACT

1. **Wiring Standard**—All workmanship and materials are to comply with the requirements of the National Electrical Code and the applicable local ordinances and the electrical plans and specifications specifically applicable to the job.

2. **Scope of Work**—Unless specifically stated otherwise in this proposal, the scope of work covered by this proposal is limited to that work specifically covered by the electrical drawings and the electrical section of the specifications.

3. **Lighting Fixtures**—Unless it is specifically included in the electrical drawings and the electrical section of the specifications or specifically stated in this proposal, the furnishing and installing of electrical lighting fixtures and lamps is not included in this proposal.

4. **Additional Work or Changes**—Additional work or changes may be ordered in writing by the customer at any time, for which the customer agrees to pay in addition to the contract price named herein at a price agreed upon or at our regular rates for time and material work.

5. **Written Orders**—The electrical contractor shall receive written orders for all additional work or changes signed by an authorized person before proceeding with such extra work or changes.

6. **Payments for Additional Work or Changes**—Payments for additional contracts to the original contract shall be made under the same terms and conditions as are embodied in the original contract.

7. **Contract Payments**—The electrical contractor shall not be required to proceed with the installation of the work if the payments applying on same have not been made as specified in the contract.

8. **Unavoidable Interruptions**—It is hereby mutually agreed that the electrical contractor shall not be held responsible or liable for any loss, damage or delay caused by fire, strikes, civil or military authority or any other cause beyond his control.

9. **Charges to the Electrical Contractor**—The electrical contractor shall not be liable for any charges for temporary wiring, electrical energy, heat, job cleanup, hoisting, job telephone, job office or storage space, etc., unless specifically so stated in this proposal.

10. **Liquidated Damages**—The electrical contractor shall not be liable for any charges for liquidated damages resulting from delay in completion of the work caused by factors beyond his control.

11. **Transfer of Title**—If the customer disposes of the real estate by sale or otherwise, the full amount remaining unpaid on this contract becomes due at once and payable within 48 hours after date of such proposal.

12. **Arbitration**—Any controversy or claim arising out of or relating to this contract or the breach thereof, shall be settled by arbitration, in accordance with the rules, then obtaining, of the American Arbitration Association, and judgment upon the award rendered may be entered in the highest court of the forum, state or federal, having jurisdiction.

Courtesy: National Electrical Contractors Association, Inc.

Bid Form
Figure 11-7 (continued)

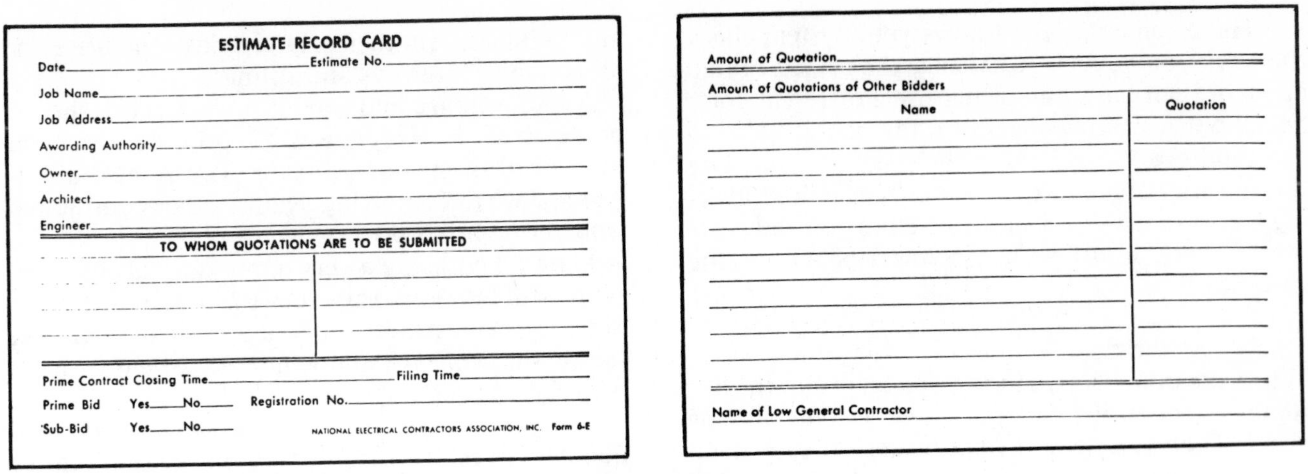

ESTIMATE RECORD CARD

Date_____ Estimate No._____

Job Name_____

Job Address_____

Awarding Authority_____

Owner_____

Architect_____

Engineer_____

TO WHOM QUOTATIONS ARE TO BE SUBMITTED

Prime Contract Closing Time_____ Filing Time_____

Prime Bid   Yes___No___   Registration No._____

Sub-Bid   Yes___No___   NATIONAL ELECTRICAL CONTRACTORS ASSOCIATION, INC. Form 6-E

Amount of Quotation_____

Amount of Quotations of Other Bidders

Name	Quotation

Name of Low General Contractor_____

Bid Record
Figure 11-8

Bid instructions usually describe the owner's rights in awarding the bid, the bid form, items that must be included and the deadline for submitting the bid. They all tell where the bid must be sent, when the bid will be opened and when the contract will be awarded.

Bid Openings and Lettings

The time and place of opening the bid is called the *opening*. Many jobs require that an officer of the company submitting a bid be present at the opening.

Anyone bidding a good-sized job for the first time should make it a point to attend a bid opening for a similar job. Become familiar with the bidding process. You'll pick up hints on how to prepare your bids.

Preparing the Estimate and Bid

Estimating and bidding is demanding work. Not everyone can become a competent estimator. Training, experience and attention to detail are essential. You need a good working knowledge of labor and material costs. You must know when, where and how to bargain. You must be a good judge of the amount of work your crews can handle in a given time. Finally, you must be able to evaluate the many subtle factors that make a job a winner or a big loser.

You get this only by becoming thoroughly familiar with all parts of the job and the bidding process. Accurate job records are a big help. Your file of job cost records can become your most valuable asset as an estimator.

Using The Take-off Sheet

Every estimate starts with a take-off sheet and a plan. Here are some basic rules to follow:

• Be sure plan sets and specifications are complete before you begin work. Be sure all addenda are included. Visit the job site if possible.

• Read the electrical specifications (Division 16) completely. Underline and note important information. For example, switch and fixture mounting heights for elderly housing are different than in homes. This may not be shown on the plans. But it will be in the specs. Notice this now, not later, when making the correction can be very expensive. Read through the General Requirements and the Equipment Divisions. Information in other divisions can affect your costs. Don't assume all of your work is outlined in the electrical section. Check plumbing, heating and architectural plans and specifications for electrical work that has to be included in your bid.

• Examine the electrical plan completely. Also look at the site plan and building elevations. While these drawings may not affect the electrical work directly, they give you a feel for the project. Pay close attention to notes and details.

• Don't assume or correct any information. Discuss errors, omissions, and doubts with the architect.

• Start at the upper left of the top plan sheet. Work from left to right and top to bottom. When the first sheet is completed, go on to the next sheet. Work carefully and logically. Beware of interruptions. As each item is listed on your take-off sheet, check it off on the plan with a colored pencil. Use the same color pencils and the same system on every plan.

• Have someone who knows your system check the work you've done. Challenge the person to find an error. But have him or her use a different color pencil when making checks on the plan.

• Find every cost item in the specifications. Use the specification index to locate what you're looking for. As each item is transferred to your take-off sheet, check it off with a colored pencil on the specifications.

Pricing Materials

When the take-off sheet is complete, group the like items by quantity and put them on the pricing sheet. Then you need to find the cost of each item. You'll need the most competitive prices if your bid is going to be competitive. Be alert for possible material substitutions at lower cost.

Some electrical contractors consider only price when buying materials. They check prices with many suppliers, even very large suppliers in distant cities. Others give as much business as possible to one local supplier, going elsewhere only when that supplier doesn't have what's required.

The best buying uses both of these methods. You want the best deal. But price isn't the only consideration. Supplier cooperation, good delivery and consistent service are also important. You'll have emergencies when a key supplier can help. If all your suppliers are faceless phone numbers in distant cities, it will be very hard to get special consideration. Use judgement when buying materials. Have one or more regular suppliers. But keep them competitive by getting quotes from other dealers, especially on larger orders.

Suppliers can help you by giving advance notice of price increases for commonly used materials. He may offer to sell existing inventory at the old price. You have to weigh the dollars saved against the cost of handling and storing the material. But that's an attractive offer if you have the cash to carry the materials until they're needed.

When a supplier quotes a price and you use that price in an estimate, get that quote in writing. A quotation is more valuable to you if it is guaranteed until a certain date. Without a guarantee, the price of the material could rise between the time the bid is accepted and material ordered. You would have to absorb the added cost. That money comes out of your profit.

Equipment and material quotations should be itemized. On large jobs, some manufacturers and suppliers will submit a *package* or *job lot* quotation. The supplier quotes only one price for the items requested. For example, one supplier may give a single price for all the panelboards, starters and switches. Another may give just one price for all conduit, wireways and fittings.

Avoid package and job lot pricing. They should be itemized. Accepting a single price deprives you of control over your estimate. You're at a disadvantage when competing against other contractors who get itemized prices.

Material costs are a very important part of every successful bid. Give this work the time and effort it deserves. You'll need large amounts of patience, tact, shrewdness and humor when pricing materials.

Labor costs are easier to calculate. Labor productivity doesn't fluctuate very much from year to year. But you have to stay posted on wage changes—including prospective increases that may affect your craftsmen.

Contingencies

After calculating the material and labor costs, add what you feel is the proper percentages for overhead and profit. On a larger job, or a job that includes some estimating uncertainty, add several percent to cover contingencies.

A contingency allowance allows for problems or costs that cannot be estimated before the project begins. Construction seldom goes smoother than planned. Nearly all surprises are bad surprises. They add to your cost and reduce your profit. For example, delays caused by bad weather or late material shipment could cause unscheduled overtime. Work estimated at straight time and actually done at time-and-a-half pay will inflate costs. This is the type of item covered by a contingency allowance.

Some contractors allow for contingencies by adding a few extra cents to low cost materials in the bid. Still others add 2 or 3% to the total bid as a contingency allowance.

Most contractors agree that it's good estimating practice to allow for contingencies when there is more than usual uncertainty in the job. The art of estimating is to identify every cost item and include it in your bid. Contingency is a legitimate cost item in this uncertain world and belongs in many bids.

Minority Rules And Regulations

Since 1966, there have been federal, and in some areas, state and local laws outlining the hiring of women and minorities on construction projects. These requirements are based on job size—both dollar value and the number of people employed. Minority rules affect subcontractors as well as general contractors. Low bidders have lost jobs because they didn't show proof of compliance with minority regulations.

Laws such as these may change over time. Much of their impact depends on how strictly they are enforced. A contractor acting in good conscience will provide job opportunities to all qualified people regardless of race, sex or other differences. But you may have to do more than that to comply for some contracts. Be aware of the laws and how they are enforced in your work area.

The Quotation

The final bid, sometimes called the quotation, is your firm and final offer to provide all the labor, materials and skill necessary to do the work described in the bid invitation and job contract. The quotation can be simple or detailed, depending on the bid instructions. A quote for a large job will usually include the following:

• The base bid and addenda covered by that bid.
• Alternative bids, if needed.
• Brief statements covering:
1. Start up and completion times
2. Work schedules
3. Material substitutions and their approvals
4. Exceptions to any items in the plans and specifications and the reasons why the contractor does not include them
5. Addenda included and not included in the bid
• Bid security, if required
• Payment schedule, if requested in the instructions
• Signature of the contractor or his legal representative

Check the figures in your quote several times for accuracy and completeness. Check instructions in the bidding invitation. Failure to follow procedures will usually disqualify your bid. Here are some points even seasoned estimators can overlook:
• Bid form. Be sure to use the bid form provided by the architect, owner or general contractor. If you are free to use your own, it's a good idea to submit a sample form for prior approval. This is especially true if this is your first submittal to this particular company.
• Bid receipt. A date, time and place for the receipt of the sealed bid will be given in the invitation. Check these instructions and follow them carefully. In some cases, bid extensions are given for good cause. But don't count on it.

Bid Errors

Bids with errors may be disqualified. Sometimes the entire bid process must be repeated. This is bad for all bidders since everyone now knows what the competition's costs are. The next letting usually results in lower bids by everyone, cutting the possible profits.

The most common sources of errors in bids are ordinary arithmetic, copying figures from one sheet to another and forgetting to transfer items from one sheet to another. These mistakes can be avoided by working slowly and carefully.

Have someone check all arithmetic. Cross out items as they are transferred or calculated. Do all calculations twice. Use the same system every time. If you are interrupted while working on the bid, finish the transfer or figure and cross it out before turning your attention to something else. Be very careful of misplaced decimal points. This is one of the most serious and most common mistakes made in bids.

Develop a checklist that shows all the cost items you can think of on the type of jobs you handle. Go over the checklist carefully before finishing the estimate. Figure 11-9 shows many of the cost items common on most electrical jobs.

Errors of judgement are the most difficult to identify and correct. All estimating involves judgement. That's what makes estimating a highly paid and skilled profession.

Failure to anticipate rising labor and material costs, failure to recognize job site conditions and overestimating crew productivity are all errors of judgement.

You can't be sure there are no errors of judgement in your estimates. But you can improve your judgement by getting the facts needed to made a good decision. Try to get material costs that are guaranteed. If you can't, see how much material costs have changed over the last 6 months or year and add that percentage change to your bid.

Anticipate changes in labor rates. Job site and crew productivity errors will be smaller if you make site inspection both before work begins and while the work is in progress. Accurate job records give you the hard facts about what your costs are.

Your judgement is needed on less obvious problems. The overall state of the economy affects the construction industry and all the contractors you bid against. The amount of work available in your work area influences the number of bids submitted for any particular job. Competition may be intense during some times of the year and almost nothing a

FORM 21E
NATIONAL ELECTRICAL CONTRACTORS ASSOCIATION, INC.

PRE-QUOTE ESTIMATE REVIEW CHECK LIST
(See Reverse Side For Recommended Use)

Job No. _____

Estimate No. _____

Job Name _____

Date _____

Job Location _____

Number of Estimate Sheets _____

Awarding Authority _____

Bid Depository No. _____ Closing Time _____ Filing Time _____

Estimated By _____ Estimate Final Approval By _____

A	DOUBLE CHECKED FOR GENERAL CONDITIONS, OMISSIONS and ERRORS

1. Scope of Work (See Reverse Side): No___ Yes___ By___ 3. All Electric Drawing Sheets Accounted For: No___ Yes___ By___
2. Correct Drawing Scale: No___ Yes___ By___ 4. Job Site Checked: No___ Yes___ By___

PRICING SHEETS

5. All Pricing Sheets Checked For:
   Omitted Unit Prices: 6. All Pricing Sheet Extensions: No___ Yes___ By___
   Omitted Lump Sum Prices: No___ Yes___ By___ 7. All Pricing Sheet Totals: No___ Yes___ By___
   Omitted Labor Units: No___ Yes___ By___ 8. All Pricing Sheet Total or Sub-Total
   Omitted Lump Sum Allowances: No___ Yes___ By___   Transfer To Summary Sheet: No___ Yes___ By___

SUMMARY SHEET

9. Summary Sheet Checked For: 13. Special Supervision – Engineering: No___ Yes___ By___
   Omitted Material Sub-totals: No___ Yes___ By___ 14. Licensing Inspection Requirements: No___ Yes___ By___
   Omitted Labor Sub-totals: No___ Yes___ By___ 15. Special Labor Agreement Requirements: No___ Yes___ By___
   Material Cost Additions: No___ Yes___ By___ 16. Prime Cost Additions: No___ Yes___ By___
   Labor Hour Addition: No___ Yes___ By___ 17. Overhead Allowance Included: No___ Yes___ By___
   Labor Cost Extension: No___ Yes___ By___ 18. Total Cost Addition: No___ Yes___ By___
10. Direct Job Expense Items: No___ Yes___ By___ 19. Profit Allowance Included: No___ Yes___ By___
11. Direct Job Expense Calculations: No___ Yes___ By___ 20. Inclusion of Necessary Sales and Excise
12. Sub-Contract Requirements: No___ Yes___ By___   Taxes, Bonds, Fire & Liability Insurance: No___ Yes___ By___
   21. Total Estimated Price Addition No___ Yes___ By___

B	CHECK POINT AVERAGES

1. Average Price Per All Type Outlets. (Compare with Similar Jobs. If Seems High or Low Ascertain Reason)
   $\frac{(Total\ Estimated\ Price)\ (\$_____)}{(Total\ Outlet\ Boxes\ )\ (\quad)} = \$\_\_\_\_\_$ Per Outlet

2. Average Price Per Square Foot. (Compare with Similar Jobs. If Seems High or Low Ascertain Reason)
   $\frac{(Total\ Estimated\ Price)\ (\$\_\_\_\_\_)}{(Total\ Floor\ Area\ )\ (\quad Sq.\ Ft)} = \$\_\_\_\_\_$ Per Sq. Ft.

3. Average Branch Circuit Run Length (If Appears To Be Too Low or High Check Drawing Scale Used)
   $\frac{(Total\ Conduit - EMT\ \frac{1}{2}"\ to\ 1")\ (\_\_\_\_\_Ft)}{(Total\ Outlet\ Boxes\ \&\ Fittings)\ (\quad)} = \_\_\_\_\_$ Ft. Per Run

4. Average Number Branch Circuit Conductors (#18-8). (If Less Than Two Check Wire Calculations & Quantity MT Raceway.
   $\frac{(Total\ Wire\ \#18 - \#8)\ (\_\_\_\_\_Ft)}{(Total\ Raceway\ \frac{1}{2}"\ To\ 1")(\quad Ft)} = \_\_\_\_\_$ Wires Per Run

5. Average Number Feeder Size Conductors (#6 & up). (If Less Than Three Check Wire Calculations & Quantity MT Raceway.
   $\frac{(Total\ Wire\ \#6\ \&\ Larger)\ (\_\_\_\_\_Ft)}{(Total\ Raceway\ 1\frac{1}{4}"\ \&\ Larger)\ (\quad Ft)} = \_\_\_\_\_$ Wires Per Run

C	DEPARTMENTAL REVIEW and APPROVAL

CONSTRUCTION 1. COMPLETION TIME: Prime Contract Period – Calendar Days _____ Weeks _____ Electrical Work Weeks _____

2. ELECTRICAL COMPLETION TIME: $\frac{(Total\ Estimated\ Man\ Hours\ \_\_\_\_\_)}{(Work\ Week\ Hours) \times (Average\ Number\ Workers)\ (\quad) \times (\quad)}$ = Estimated Work Weeks _____

3. Possible Complete Within Contract Period: No___ Yes___ By___. 4. Possible Complete Within Estimate Man Hours: No___ Yes___ By___
5. Any Special Situations Involved: Labor___ Materials___ Tools___ Subcontracts___ Licensing___ Inspection___: Explain _____

PURCHASING 1. Anticipate Material & Equipment Available To Maintain Progress Schedule: No___ Yes___ By___
2. Can Purchase Materials & Equipment Within Estimated Prices Used: No___ Yes___ By___

FINANCIAL 1. OVERHEAD: Firm's Average ___% (Prime Cost) (Labor): Estimate Overhead _____ % (Prime Cost) (Labor)
2. Estimate Overhead Allowance Adequate: No___ Yes___ By___ : Explain _____
3. Any Special Financial Problems Involved: No___ Yes___ By___ : Explain _____

4. Customer Credit OK: No___ Yes___ By___ Basis of Payment: O.K.: No___ Yes___ By_____
5. Satisfactory Profit Anticipated: No Yes By : Explain _____

FINAL APPROVAL $_____ Estimate Price Approved for Bidding: No___ Yes___ By_____
Comments:

Estimate Check List
Figure 11-9

RECOMMENDATIONS FOR USE

1. The use of this form should be started in the estimating department at the start of estimate preparation.
2. When any given phase of the estimate is completed and double checked it should be so noted on the front of this form. Do not ever check an item as "Yes" unless it has been double checked by the estimator or someone else. If sufficient time has not been available to double check any item, check it as "No". Doing so visually indicates to the person making the final review the extent to which the estimate has or has not been double checked. A decision is made whether bid the job based upon only a partial estimate "check-out".
3. The DEPARTMENTAL REVIEW AND APPROVAL section is an important feature of the use of this form. This additional "check-out" by the different responsible parties serves as an additional safeguard for a complete and proper estimate and bid. Responsibility for final approval is clearly indicated.
4. The section of the form below is normally filled out by the estimator during initial review of the contract specifications and drawings. The same form section is included as part of NECA Form 12 EF ESTIMATE SUMMARY. When that form is filled out it will serve in lieu of filling out this section below.

SCOPE OF WORK and SPECIFICATION-DRAWING CHECK SHEET

Owner _____ Architect _____ Engineer _____

Prime Bid _____ Subbid _____ Lump Sum _____ Alternates _____

Completion Time _____ Liquidated Damages _____

Method of Payment _____ Progress Payment Retention _____

Number of Electrical Drawings _____ Drawing Scale _____

Electrical Work Shown on Architectural _____ Mechanical _____ Drawings

Number and Dates of Addenda _____

Electrical Work Laid Out: Yes _____ No _____ All Electrical Work Shown Included in Contract: Yes _____ No _____

All Material and Equipment Furnished by Contractor: Yes _____ No _____ Material by Others: All _____ Part _____

Explain _____

Fixtures Supplied By _____ Installed By _____ Lamps By _____
Signal Equipment By _____ Motor Starters By _____ Controls By _____
Power Equipment By _____ Vault Equipment By _____ Switchgear By _____
Painting By _____ Clean-up By _____ Patching By _____
Trenching By _____ Duct Work By _____ Concrete Work By _____
Temporary Power Wiring By _____ Extent of _____

Electrical Work in Other Sections of Specifications _____

Number of Buildings _____ Number of Floors _____ Total Floor Area _____
Total Lighting Load _____ Total Power Load _____
Branch Circuit Wiring Method: Frame or Suspended Ceiling _____ Slab _____ Exposed _____
Feeder Wiring Method: Frame or Suspended Ceiling _____ Slab _____ Exposed _____
Service Wiring Method: _____ Overhead _____ Underground _____
Switchboards: Enclosed switches on _____ Metal Enclosed _____ Switch and Fuse _____ Circuit Breaker _____
Panel Boards: Fused _____ Circuit Breaker _____ Type _____
Point of Connection to Utility Supply _____
Other Data _____

Type of Job	*Type of Building Structure*	*Type of Construction*
Dwelling _____	Wood Frame _____	Distance between Floors _____
Multi-Family _____	Wood Truss _____	Attic Above _____ Suspended Ceilings _____
Commercial _____	Masonry _____	Basement Below _____ Slab on Ground _____
Institutional _____	Reinforced Concrete _____	Ceilings: Plastered _____ Acoustic Tile _____
Industrial _____	Structural Steel _____	Partitions: Frame___ Masonry ___Concrete _____
Warehouse _____	Metal Decking _____	Special Partitions: _____
Recreational _____	Cellular Floor _____	Walls: Concrete Pour___Tilt-up___Masonry _____
Power Plant _____	Open Steel Frame _____	Special Walls: _____
Substation _____	Prefabricated _____	Slab: Flat__Pan-Joist__Lift__Pre-cast _____
Other_____	Other_____	Slab Pour: Single _____ Separate Rough & Finish__
		Other: _____

Architectural Drawings Checked: Date _____ . Mechanical Drawings Checked: Date _____
Specification General Conditions Checked: Date _____ . Mechanical Specification Checked: Date _____
Insurance Requirements Checked: Date _____ Job Site Checked: Date _____

_____
(Estimator's Signature)

Estimate Check List
Figure 11-9 (continued)

few months later. Traditionally, bids let in spring are more competitive. Some jobs go begging later in the season when work is in full swing.

Use all the information available to you. Construction magazines and reporting services such as the Dodge Report list jobs going to bid and work still in the design process. Your suppliers and your banker are good sources of information about local work.

Bidding Methods

This section describes two methods of bidding widely used in the construction industry. As you learn more about bidding and about your business, you may want to adjust them to detail more accurately the work you hope to get.

The simplest way of figuring a bid is the *lump sum* method: On most jobs the cost of materials will be about 60% of the total job cost, and labor cost will be about 40%.

Assume that your estimate shows that materials for a job will cost $6,000. Adding your labor costs to materials costs will give you your base cost. First find the base cost:

$$\frac{\$6,000 \text{ (material cost)}}{.60 \text{ (percent material cost)}} = \$10,000 \text{ base cost}$$

To get the labor cost:
$10,000 (base cost) x .40 (percent labor cost)
$$= \$4,000$$

After you get your base cost, you add your overhead and profit to get the total bid. Your final breakdown will look like this:

Materials	—	=	$6,000
Labor	—	=	4,000
Base Cost	—		10,000
Overhead	— ($10,000 x 25%)	=	2,500
Profit	— ($10,000 x 10%)	=	1,000
Total Bid		=	$13,500

Be sure to take 40% times the base cost and not the material cost to get the right labor cost. If you want to find the manhours needed for the job, divide the labor cost by your hourly labor cost.

The 60-40 lump sum bid is also handy for doing a quick check of estimates. Naturally, this will vary with the type of job, labor scale and crew productivity. But estimating material costs is fairly easy and can be reasonably accurate. Once the material cost is known, the estimate total should be about

2.25 times the material cost. An estimate with a much higher or lower total may still be right because of special job needs. But it alerts you to a possible mistake if your estimating for a "typical" job varies widely from this standard.

On larger jobs, you will need more detailed estimates. A detailed estimate should give you a complete picture of job costs. You can then adjust your bid in areas where your company can cut costs, such as crew productivity.

A more detailed estimate begins, like the lump sum method, with a complete material list on your take-off sheet. The material prices are then added in the usual way on the price sheet. Then installation manhours are established for each of the materials to be installed. The total manhours times the wage cost gives the labor cost.

A detailed cost estimate can take a lot of time on a large job. But there's a good way to speed up the process: the *unit method.* This system groups materials that are usually installed together as a unit and calculates a selling price for that unit. The labor can be given in either manhours or in dollars, as shown in Figure 11-10. Summarize labor and material costs at the bottom of the worksheet and add tax, overhead and profit. Include a sketch to make identification of the unit easier. The total is your selling price for the unit.

Make up a unit worksheet for the material groups you handle most. Arrange the sheets in a logical order in a looseleaf notebook. This unit price book will become your most useful and frequently-used estimating reference if you keep it current. Add new items as needed and pencil in changes to prices and manhours when they take place.

Figure 11-11 shows how the receptacle outlet in Figure 11-10 is priced with both the conventional method and the unit method.

Cost Management and Labor Supervision

Preparing an accurate estimate and bid will help get you the job. But that's only getting started. To make a profit from the work, you have to see that the job is done the way you bid it. Actual labor and material costs have to reflect your estimated costs as closely as possible. Of course, if you can do the job for less than estimated, you increase your profit.

Protect your material costs by doing the following:

• Get quoted prices in writing, for a specific length of time.

• Check all invoices and bills to make sure your billed price is the same as the quoted one.

Date: **10-23-81** Item: ⊖ w/ *gdg.* + NM Cable Page____of____Pages

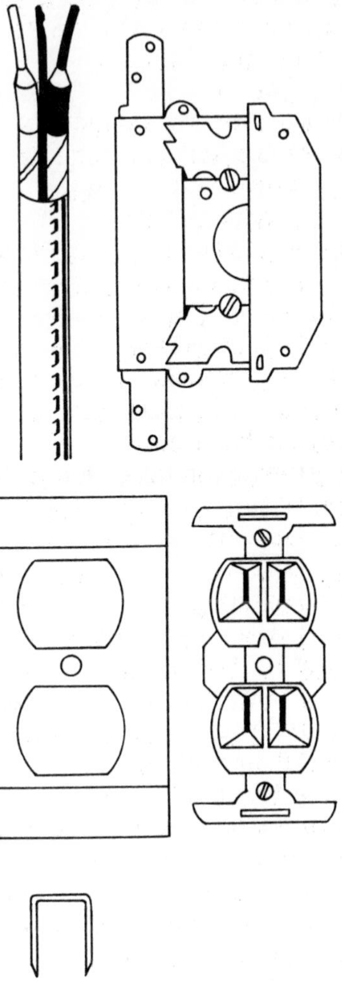

No. Description		Material		Labor	
1 Nail Box			.30		.20
1 Duplex Recp.-gdg			.25		.30
Staples			.03		-
1 Plate, plastic			.15		-
#12-2G NM Cable	20'x.06	1.20	.06/ft.		1.20
Totals		1 1.93		4 1.70	
		2		5	
		3		6	

		1	2	3
Material		1.93		
Sales Tax	4%	.08		
Labor		4 1.70	5	6
P.R. Tax	15%	.26		
Job Cost		3.97		
Overhead	25%	1.00		
Profit	10%	.40		
Selling Price		5.37		

Unit Method Worksheet
Figure 11-10

• Insure and lock-up materials stored on the job site. Protect yourself against material loss due to damage, theft or weather by having a fenced-off area, storage shed or trailer. Large equipment not easily stored inside should be kept away from traffic areas where it might get damaged by the other trades. Even though it won't be your fault, you'll have the delay and problem of replacing it.

• Have an inventory system that's kept up-to-date. At any point of the job, you should know what equipment has been installed, what's been ordered but not delivered and what's in storage awaiting installation.

Increasing labor productivity goes a long way towards increasing your profit. The best way to increase productivity is through good supervision. Treat your men fairly but expect a full day's work from them. Choose your foreman or supervisor carefully. He must, of course, get along with the crew. But he shouldn't be too friendly, either. Don't be afraid to push your workers if the job demands it. But use tact and understanding. Enlist their help instead of bullying them. When the job is done, reward those who did the work and have shown their loyalty to you. A Christmas bonus, a night out with the wife or girlfriend, a word of

GENERAL ESTIMATE

BUILDING _____

LOCATION _____

ARCHITECTS _____

SUBJECT _____

ESTIMATE NO. _____

SHEET NO. _____

ESTIMATOR _____

CHECKER _____

DATE _____

DESCRIPTION OF WORK	NO. OF PIECES	DIMENSIONS		EXTENSIONS	EXTENSIONS	TOTAL ESTIMATED QUANTITY	UNIT PRICE M'T'L	TOTAL ESTIMATED MATERIAL COST	UNIT PRICE L'B'R	TOTAL ESTIMATED LABOR COST
CONVENTIONAL METHOD										
Nail Box						4	.30	1 20	.20	80
Dup. Recep. gdg						4	.25	1 00	.30	1 20
Staples						—	.03	12	—	—
Plate, plastic						4	.15	60	—	—
#12-2 G NM Cable - 20'						4	1.20	4 80	.06/ft.	4 80
Total							1.93	7 72		6 80
UNIT METHOD										
Dup. ⊖ gdg						4	1.93	7 72	1.70	6 80

Conventional Method Compared
With Unit Method
Figure 11-11

praise for a difficult job well-done, given at the right time, will do much to insure your success on future jobs.

When your company starts to get a large number of jobs all going at the same time, keep a close watch on all of them through daily reports, either written or over the phone. Have a weekly summary comparing job progress with labor costs on your desk by Monday afternoon. If a job is reported 30% done but the labor cost is already at 50% of the total estimated, you've got a problem. Move quickly to correct it.

Low productivity and high material costs will ruin your company. Avoiding them will be a major part of your job once your company is underway.

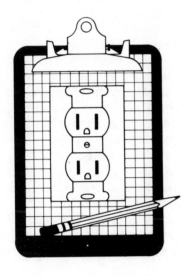

Chapter 12:

Staying in Business

Every year many new electrical contractors go into business for themselves. Many will give up and go back to working for someone else within a year or two. Others will grow and become established firms within a relatively short time. Why do some succeed while others don't?

There are many reasons for failure in the electrical contracting business. Inadequate capital, poor bookkeeping and lack of work are the most common reasons mentioned. However, any single reason seldom tells the whole story.

The real reason for failure is usually that the contractor wasn't able to put together all parts of the puzzle. He may have had the capital but overestimated the amount of work available. Perhaps the work was there but he wasn't able to master the fundamentals of running a business—payroll, taxes, bookkeeping, estimating, planning and sales.

The reason for failure is seldom that the contractor wasn't technically qualified to do or to supervise the work. There's one quality most new electrical contractors share: They're very good electricians. That's why they want to go into business on their own.

But when you're just getting started, it takes only one or two mistakes in critical areas to put you out of business. And these critical mistakes won't be made on the job by an electrician. They'll be made by the contractor and they'll be about business operations, not wiring details.

A good way to avoid these mistakes is to test your business knowledge and strengthen areas where you may be weak. Review the last two chapters. Make a list of the topics and quiz yourself. Do you have the skills and knowledge that electrical contractors must have? Being prepared to run an electrical contracting business can be one of your important strengths.

If you don't like to deal with taxes, or bookkeeping, or estimating, or some other subject we've covered, it's probably because you lack the necessary skills. Concentrate on these subjects. Review the information in this manual. Go to the library and dig out the necessary information. Keep in mind that you have to be ready to wear two caps—electrician and business manager—and be equally comfortable with both:

• No one but you can master every detail of your business.

• Keep a good set of books. Keep them up to date.

• Be aggressive but fair with customers and employees.

• Value your reputation for honesty and competence.

• Deal with those who value their reputation for honesty and competence.

• Working capital is the lifeblood of your business. Protect it.

• Be ready to work harder and longer than anyone on your payroll.

Interviews With Successful Contractors

No one can guarantee that your business will succeed. But studying the success of others may help. The rest of this chapter consists of three interviews with successful electrical contractors. All three companies have been in business for at least 20 years. They operate in cities of various sizes. All began as a small or a one-person operation. The names used are not the real names. But these companies do exist and probably will continue to thrive so long as a hard-working, experienced, knowledgeable electrical contractor is available to run them.

Standard Electric Company

Bill M. is the president of Standard Electric. It's been in business since 1912, but Bill has been with the company for only 20 years. He worked his way up from electrician to his present position. Located in a city of about 150,000 people, the company has had an annual sales volume between $800,000 and $900,000 in recent years. During peak times, it employs 10 to 14 people.

Bookkeeping - "In the beginning the system was pretty simple. We did it here at the office and once a year an accountant came in to do the balance sheet and taxes. As the volume grew and the government demanded more paperwork, we went to having him do it quarterly. Now we're at it once a month. The daily bookkeeping is still done here at the office with the system we've evolved through the years."

Financing - "The company has always financed through one bank, at least since I've been with it. It's a good working relationship. I've never had reason to change it. Over the years we've always been able to get both long- and short-term financing at pretty reasonable rates."

Getting Work - "We get most of our work from the Dodge Report, bid invitations or bid notices in the newspapers. We've been around long enough so the general contractors and architects in the area know about us. Once in a while a call comes in from our ad in the yellow pages."

Bid Preparation - "Be sure to organize. That's the key. Keep track of your sheets and don't leave anything out. It takes a lot of experience to do it right and I just don't know how you get it except to go out and do it. The work you're most familiar with is the easiest to bid. Work that you haven't bid on before—it seems you're either high or low. The biggest pitfall is the other guy. You prepare a good bid that you know is close—say around $9,000 or $10,000. He's way off base. He's going to lose his shirt, but he'll still get the work."

Controlling Labor Cost - "A good foreman is the answer. And you have to make a lot of visits to the site and keep close supervision. And if you're running the company, run it. The only time you should be doing field work is if you're just a small

one- or two-man operation. Or else if your back is really to the wall. But basically, your place is in the office or inspecting."

Dealing with Economic Shifts - "When things get slow, we do more advertising. We tighten our belts and go after work we normally wouldn't do, maybe smaller jobs or remodeling. We've also diversified into solar systems. If there's no electrical work, a man can go out and install the system. I had one put in my house first, though. I won't handle anything unless I know for sure it works. It's too much grief otherwise."

Advice When Starting a Business - "Have plenty of field experience behind you. Or else someone who does have it should be working for you. Learn all you can about the business end. And get all the money you can lay your hands on to carry you over your accounts receivable."

Acme Electric

Acme is located in a city of 300,000 people. It was started in 1960 by Jake K. and for the first two years was a one-man operation. Gross sales volume was about $40,000 annually. In 1962 business grew enough for Jake to hire his first employee, a young electrician named Jerry. He's now the company president. Jake is still active in some company operations. The two men are jointly responsible for Acme's annual sales volume of between $750,000 and $800,000. There are usually about 12 people on the payroll.

At one time the bulk of Acme's work was residential. Emphasis has been more on commercial in recent years.

Bookkeeping - "The first ten years or so I did the books myself—payroll, billing, the whole works," Jakes says. "But it got to be too much. The accounts receivable began to run further and further behind. Now we do the daily work here in the office. Once a month it's sent to a computer company for payroll and posting. Then to our accountant for review."

Dealing With Economic Shifts - "This business is either feast or famine," Jerry says. "You've got to set enough aside during the good times to get through the bad. Sometimes too, if the work isn't where you normally get it, you have to bid on jobs you maybe wouldn't be interested in. But both Jake and myself agree, we won't bid if we don't make money. That idea of getting a job just to keep the men busy—all you do is end up trading dollars."

Advertising - "We've done some but not a lot," Jake says. "We've never had to. The word-of-mouth of the craftsmanship of our jobs has always been enough. Our best recommendations come from the building inspectors who have checked our work."

Getting Work - "We don't get much from the Dodge Report," Jake explains. "We do bid but it's with individual contractors, mostly those we're familiar with. Over the years we've purged our list of contractors that we'll work with. Some cost me a lot of money when they went broke. Others were just out to get the cheapest job possible. I quit doing business with them. Most of our work is repeat work with builders and contractors we know."

Bid Preparation - Both Jake and Jerry recall making mistakes on a bid and having to live with the results.

"Once I missed a whole living room with cove lighting and 15 or 16 outlets," Jake remembers. "When I went back to the contractor, he said, 'That's tough.' It took the profit right out of the job."

Fortunately, such mistakes were few. The best way to avoid them, both men agree, is to "have more than one person double-check the bid—both the take-off and the figures."

Controlling Labor Cost - "It's up to your foreman or supervisor to push the job through," Jerry points out. "They have to make sure the work gets done on schedule. Your part is to give them what they need. The tricky thing is material cost. There was a period when we bid a lot of work and then material costs sky-rocketed. Service gear and things like that we could get a fixed cost on. But materials like wire and other copper went crazy. We lost our shirt on some jobs. Still, we managed to keep going.

Financing - Jake ran Acme as a sole proprietorship from 1960 to 1965 and did his own financing. In 1965 he incorporated and refinanced, using his house as collateral for the loan. The capital was used to expand operations. Since that time it's been, "Us and the bank."

"We've dealt with the bank ever since. But lately they were bought out by a bigger bank. Now they seem only interested in charging all they can get for their money. They don't seem to value the past relationship. We're considering going somewhere else."

Advice When Starting a Business - "Make sure you have a good amount of work," Jerry advises. "Then the most important thing to watch is your billing and cash flow. A lot of people go into business and offer extended credit terms and expect to get paid eventually. They just take the money as it dwindles in. Nowadays you can't do that. If you've got receivables over 30 days, you have to go out and get them. You can't carry accounts like you could five or six years ago."

County-Wide Electric

Harold W. bought County-Wide Electric in 1956. He had worked for the company for several years before making the purchase. At that time the company had only two other employees. Harold feels this was a good way to get started, since he had already dealt with some of the customers. He had also established a good reputation for the quality of his work and had plenty of field experience. Still, he remembers the first 10 years of operation as being "plenty tough."

When he took over, the company did mostly local residential work. Harold expanded operations into light commercial work over a county-wide area. Later, he got a contract with the telephone company to wire relay stations in many parts of the state—work that made long travel times necessary. He would recommend this type of work only for an established company. "You've got to have someone in charge you know can do the work. Otherwise, you'll run yourself ragged trying to keep track of work that's all over the state."

Currently Harold's company employs about a dozen men. Sales volume averages about $750,000 annually.

Licensing - "Each town, village and city we go into has its own licensing and inspector. We tried a couple of times to get a state license established but so far, no luck. So now for every little town or village, we have to go in and get a license. Most of the time when they see we're already licensed by a big city they just take the fee and issue a permit. That's all it really is, a lot of times—a way for these small towns to pick up some extra revenue. Still, it's a lot of trouble for us. But it's important to get along with the inspector who has jurisdiction."

Financing - "I had my house paid for when I went to the bank to take over the business. So it was no problem to borrow. I figure I've paid for my house three times over with borrowing on it for capital. Up until a year ago I always dealt with the same bank. But then they gave me a raw deal so I switched. If they didn't want to rely on old, valued customers, then I didn't want to rely on them."

Dealing with Economic Shifts - "You've always got to watch expenses and accounts receivable, whether it's a good year or bad. And through the years you learn who to work for and who to stay away from. What almost killed my business in one of my first years were four bankruptcies by general contractors. They cost me a bundle. Our work, our time, our money and materials were in those jobs. But what could I do? It was gone. Now most of my customers are repeat business so I don't have to worry as much. Still, I keep an eye on who we work for. They have to be reliable."

Getting Work - "We quit the Dodge Report. It got too scattered for our needs. We get a lot of bid information from Western Builder magazine. But the biggest share of our work comes from contractors who've heard of us. They come in and drop off the plans and say 'Give me a bid on this.' Some of our regular customers don't even get other bids. They know from experience that I'll do my best to be fair. That's why it's important to build a good reputation."

Bid Preparation - "If there's a fool-proof way to make up a bid, I'd like to know it. All you can do is go by the plans and specifications. You have to get material prices every time from the supplier. Most of them you can get a 30-day hold on (the price) but that's about it. Before you submit the bid, go over it at least twice to make sure you haven't missed anything. Sometimes if a lot of time has passed between figuring and the submission, you have to check back with your suppliers on prices."

Controlling Labor Cost - "Labor cost stays pretty well constant because of wage contracts. But you have keep checking the jobs to be sure the work is getting done. Labor overruns are a problem only if you don't have good help. I haven't always been able to get good men, but I've tried to keep the best I have. That way, over time, I've built up a good crew."

Bookkeeping - "We do the daily bookkeeping here at the office. We just bought a little personal computer that should cut down on the time. Once a month the accountant goes over the records and closes the books for üs."

Advertising - "I've only run an ad in the yellow pages of the phone book. We rely on reputation and word-of-mouth. It must be working. I've managed to stay busy for almost 30 years.

Advice When Starting a Business - "Try and start with as much of your own money as possible. Nowadays, with interest rates, even the best rate will eat into your profit. And if you're starting out cold, try and start in an area where people know you and your work. Then work on building your reputation. Be honest and reliable. When things get tough, that's what people look for. If you're only in it for a quick dollar, you won't be around long."

Index

Practical References For Builders

Estimating Electrical Construction

A practical approach to estimating materials and labor for residential and commercial electrical construction. Written by the A.S.P.E. National Estimator of the year, it explains how to use labor units, the plan take-off and the bid summary to establish an accurate estimate. Covers dealing with suppliers, pricing sheets, and how to modify labor units. Provides extensive labor unit tables, and blank forms for use on estimating your next electrical job. **272 pages, 8½ x 11, $19.00**

Residential Electrical Design

Explains what every builder needs to know about designing electrical systems for residential construction. Shows how to draw up an electrical plan from the blueprints, including the service entrance, grounding, lighting requirements for kitchen, bedroom and bath and how to lay them out. Explains how to plan electrical heating systems and what equipment you'll need, how to plan outdoor lighting, and much more. If you are a builder who ever has to plan an electrical system, you should have this book. **194 pages, 8½ x 11, $11.50**

Electrical Blueprint Reading

Shows how to read and interpret electrical drawings, wiring diagrams and specifications for construction of electrical systems in buildings. Shows how a typical lighting plan and power layout would appear on the plans and explains what the contractor would do to execute this plan. Describes how to use a panelboard or heating schedule and includes typical electrical specifications. **128 pages, 8½ x 11, $7.25**

Finish Carpentry

The time-saving methods and proven shortcuts you need to do first class finish work on any job: cornices and rakes, gutters and downspouts, wood shingle roofing, asphalt, asbestos and built-up roofing, prefabricated windows, door bucks and frames, door trim, siding, wallboard, lath and plaster, stairs and railings, cabinets, joinery, and wood flooring. **192 pages, 8½ x 11, $8.75**

Building Cost Manual

Square foot costs for residential, commercial, industrial, and farm buildings. In a few minutes you work up a reliable budget estimate based on the actual materials and design features, area, shape, wall height, number of floors and support requirements. Most important, you include all the important variables that can make any building unique from a cost standpoint. **240 pages, 8½ x 11, $10.00. Revised annually**

National Repair And Remodeling Estimator

The complete pricing guide for dwelling reconstruction costs. Reliable, specific data you can apply on every remodeling job. Up-to-date material costs and labor figures based on thousands of repair and remodeling jobs across the country. Professional estimating techniques to help determine the material needed, the quantity to order, the labor required, the correct crew size and the actual labor cost for your area. **216 pages, 8½ x 11, $15.25. Revised annually**

Estimating Home Building Costs

Estimate every phase of residential construction from site costs to the profit margin you should include in your bid. Shows how to keep track of manhours and make accurate labor cost estimates for footings, foundations, framing and sheathing finishes, electrical, plumbing and more. Explains the work being estimated and provides sample cost estimate worksheets with complete instructions for each job phase. **320 pages, 5½ x 8½, $14.00**

Contractor's Guide To The Building Code

Explains in plain English exactly what the Uniform Building Code requires and shows how to design and construct residential and light commercial buildings that will pass inspection the first time. Suggests how to work with the inspector to minimize construction costs, what common building short cuts are likely to be cited, and where exceptions are granted. If you've ever had a problem with the code or tried to make sense of the Uniform Code Book, you'll appreciate this essential reference. **304 pages, 5½ x 8½, $16.25**

Construction Superintending

Explains what the "super" should do during every job phase from taking bids to project completion on both heavy and light construction: excavation, foundations, pilings, steelwork, concrete and masonry, carpentry, plumbing, and electrical. Explains scheduling, preparing estimates, record keeping, dealing with subcontractors, and change orders. Includes the charts, forms, and established guidelines every superintendent needs. **240 pages, 8½ x 11, $22.00**

Wood Frame House Construction

From the layout of the outer walls, excavation and formwork, to finish carpentry, and painting, every step of construction is covered in detail with clear illustrations and explanations. Everything the builder needs to know about framing, roofing, siding, insulation and vapor barrier, interior finishing, floor coverings, and stairs. . .complete step by step "how to" information on what goes into building a frame house. **240 pages, 8½ x 11, $8.25. Revised edition**

National Construction Estimator

Current building costs in dollars and cents for residential, commercial and industrial construction. Prices for every commonly used building material, and the proper labor cost associated with installation of the material. Everything figured out to give you the "in place" cost in seconds. Many time-saving rules of thumb, waste and coverage factors and estimating tables are included. **416 pages, 8½ x 11, $14.75. Revised annually. Monthly cost updates for 280 key materials and the wages in the National Construction Estimator are available for $2 per month. Order the "National Construction Cost Newsletter" for $24 per year.**